AMERICAN ANDROIDS
CRITICAL EDITION

AMERICAN ANDROIDS
CRITICAL EDITION

PARIS TOSEN

Canada

This is a book of nonfiction.

First CreateSpace Independent Publishing Platform edition (revised),
May 2013

Copyright © 2013 by Paris Tosen

All rights reserved.

Some content previously published in *The Android Conspiracy* and *American Androids: Special Edition*.

Some editorial or grammatical errors may exist. This does not diminish the authenticity of this personal work.

Discretion is advised. The author is not a medical doctor. He is not qualified to treat any medical disorder. The content of this book is entirely his opinion based on his own personal experience, observation and analysis. Misuse of this information is not recommended.

ISBN 978-1-479326402

www.tosen.ca

Book design by Paris Tosen

CONTENTS

COMMENTARY

APPENDIX

Joe Lieberman   Nancy Pelosi   Max Baucus

"By the time that people started to accept my hypothesis, that we are under an android administration, we would have all been vaccinated with gene-stopping vaccines and microchipped and hooked into some quasi-mind altering network interface. In one or two years, there could be immense societal shifts as a result of more technological activation, including the implementation of round-the-clock drone observation units across all parts of the nation. On the one hand, we cannot afford the time it takes to inform the public that there are artificial people in control, but on the other hand, and if history teaches us anything, people need time to process and absorb this kind of ultra-fantastical thinking, even if it is true."

~ Paris Tosen

PREFACE

What has been documented in this book is the first thorough attempt to describe the presence of living androids and to position them within the context of an age-old agenda to rule the world. The presentation is in no way complete and included assumptions are not perfect. This is a risky undertaking that proceeds to understand a level of science that is at least 1,000 years ahead of human science. But where real scientists and anthropologists would dare not go—unless they were writing a novel—the great thinkers of the world haven't been able to discover the advanced human-looking androids on prime time US television. It is as an eclectic observer that I, in my own way, have rooted my discussion, analysis, and interpretations. It would be illogical for me to suggest that I have made this discovery because of my vast intelligence and resources, quite the contrary, I am not the most intelligent, or smart for that matter, person on the planet. Further, I have no scientific training outside of high school. I am not a neurologist and I am not an expert on artificial intelligence. But I believe that I do possess one powerful skill that makes me the perfect candidate for the task—intuition.

If you approach android theory as a sceptic you will deny yourself the ability to see through the existential illusion; you will find yourself filing this topic alongside the invaluable discussions commonly found in mental institutions or to put it alongside the Book of Revelation.

As soon as you deny and disbelieve that the illusion is fake, the illusion becomes real and then continues to be real. The reasons I have been able to succeed where all others have failed are because I am a man with a good amount of naivety and curiosity, two character traits that fit well with my intuition, but do not enhance my level of intelligence. But by making an observation without judgment and by seeing something without imprinting my beliefs onto it I have been able to pierce the existential illusion and to see it for what it is. Had I the impeccable mind of an investigative journalist or the logic of a conventional scientist, I would have failed on all counts and this discovery would have been lost.

This is the five-year anniversary of the android discovery. I did not release any of my findings for the first 23 months and even then I decided from the start to downplay the identities of the three androids because I wanted to focus on their artificial characteristics. I felt that the story should focus on these synthetic beings and that their programmed identities were not as important. For over two years I downplayed the identities of the three individuals and only reluctantly wrote their names after several book editions, the video documentaries *Androids Among Us* and *American Androids* could not help but to detail their faces. This is the first book that not only highlights the people's identities and names, but also their images.

The historical importance of this investigation, although currently understood as a phenomenon without comparison, is unknown at this time. It can be compared to finding a laptop computer in the tomb of an Egyptian Pharaoh. This is a good example of the magnitude of this discovery only that the androids are not so obviously itemized because they look like your neighbor, your parents, and your teachers. For all intents and purposes they look human. This has been a fundamental problem with the presentation of facts. I have made my best efforts to highlight the most obvious ticks and nuances that I think reveal their artificiality. These characteristics that I have highlighted are not exhaustive and certainly do not fit the portrait of other androids, but they do provide a solid platform from which to hold a reasonable discussion. This is also the other thing. The three androids in my primary research are three of many, probably in the hundreds, if not thousands.

We are fortunate to have such fine examples before us because the other synthetic humans are not so easily observed. This is why I have said that leaving scepticism and disbelief at the door is one way to guarantee that you continue to understand the android infiltration on the planet. I have till now, using various formats, identified at least a dozen other android types, more advanced than the initial three, having more subtle nuances and ticks. I did so to highlight the fact that there are more than three and that they come in all shapes and sizes. The last thing I wanted was for people to think of these three American politicians as all there is and ever will be.

The description of androids is not necessarily in line with what we have come to see as artificial beings. For the artist and the writer in their descriptions of androids, they have been limited in their interpretation of an advanced science and have therefore explained an android using mechanical engineering and robotics, but I was never obliged to rely on the traditional description of a genetically-engineered being.

Now, androids are walking and breathing creative masterpieces—all at once the culmination of a true genius at work. The creation of an android is no happenstance. They are designed from the genetic blueprint with care and attention. I have tended to lean away from the artistic beauty since I felt that it would take away from the immensity of this event, and, more importantly, the androids have been employed to steer the nation of America into a course which is not in the best interest of American citizens. This is the unfortunate thing, this deceptive circumstance that is the context of my discovery. I only discovered them as a result of an emergency repositioning of America during the financial rescue fiasco in September 2008. But without that misstep on their part, I would not have noticed them, and we would not be having this provocative discussion. This work, as detailed as it is, should be considered an introduction to a new field of *android theory*. I have completed this book on my own without any outside expert or assistance, to this effect I am responsible for its shortcomings. I would like to dedicate this book to the future generations, for it is their inspiration that has given me the strength to endure the travesties and challenges of this kind of work.

Let it be said, even if in brief, these advanced races who built the androids have well been aware of my ongoing research into their secretive strategies to impede human progress, as such they have interfered, as you can imagine, with my life on levels of imagination that are hard to describe but in which some ideas may be found in this book (or in my memoirs). Humanity has the opportunity, unlike ever before, to ascend its current level of what can best be described as 'primitive science', and if some of these ideas can be understood, we will see life in a whole new light. If the android makers are 1,000 years ahead of us then if we can learn their science we can potentially learn 1,000 years of science without having to live through 1,000 years of hardship. And that would be a wise investment.

~ *Paris Tosen*

October 2012

INTRODUCTION

This book has been written so as to explain to all American citizens of the artificial control mechanisms hidden within their government, and behind the voices and promises of the President and Vice President. The discussion is specifically designed to enlighten American citizens on the human-looking android politicians who have, along with other players, hijacked the federal constitutional republic without detection. Even as this latest edition is published and made available, America remains as the world's most powerful nation (economic, political, cultural)—China is poised to overtake its economic lead by about 2016—with the world's most powerful military (40% of global military spending), tempting treasures for any interplanetary thieves. If you arrived here from another planetary system, and you were skilled in planetary takeovers, wouldn't it make sense to hijack the most powerful nation with the largest number of resources? Whereas Germany made a formidable base of operations prior to 1939, and indeed could have been their temporary headquarters, the creation of a new series of synthetic citizens made sense as they relocated their headquarters to the United States.

If the US government continues as a nation hijacked by these technocratic masters, America will increasingly be used as a weapon to alter the geopolitical landscape; and with the adoption of new national security protocols, a sequestered American power could soon turn against its citizens and declare itself a police state (totalitarianism) ending all liberty and freedoms first adopted in the 1787 Constitution.

This book builds upon a number of handmade video documentaries and online discourses. It more fully explores the complex agenda of the android makers and their three key political marionettes: Joe Lieberman, Max Baucus, and Nancy Pelosi; seemingly real people with real jobs and real families on the surface, but on another level of perception quite entirely something different. These three persons, highlighted as my primary evidence in the investigation, in my opinion, are synthetic persons that have been tailored for their tasks.

In presenting this expansive discovery to the public at large, in as much detail as I can understand and knowing full well that my hypothesis and conclusion might be wrong; I hope to equalize the playing field by exposing an advanced criminal group of reality hijackers. This is the first time, to my knowledge, that this has ever been explained in a nonfiction book, a testament to the illusory strengths of our magical masters. Additionally, because I am dealing with extravagant ideas and multidimensional observations, the discussion cannot help but to stretch itself, as and when needed, outside the confines of logic and rationality. This is a multidimensional investigation suitable to an audience with some ability to set aside scepticism and disbelief in favor of deciding whether or not America is in big trouble; for if America is under the control of an invisible government then the future of the United States is at stake.

Although I am not American, I wholeheartedly am concerned that this great nation has been surreptitiously taken over by androids; intelligent simulacra that are completely controllable, programmable and likeable. My nation Canada, myself a proud Canadian, is a close neighbor with many intertwined partnerships over a great many years. Further, America has many international partnerships and national friendships that are under potential threat.

The impressive advances in robotics and artificial intelligence have fallen short of replicating the molecular perfections of the average human being. In fact, there is a very clear distinction between a fully-functioning robot and a living person, where, without a deeper understanding of genomic processes, the exactitudes of existence have remained elusive and basic requirements to the full replacement of any one of us.

The most successful aspects of this fantastical task have been relegated to the imaginative storylines of science fiction writers in the likes of Philip K. Dick and Isaac Asimov, who, although not roboticists, or artificial intelligence experts, clearly brought to surface the cultural implications of an advanced scientific process. Equipped with amazing feats of strength, stamina, and agility, the android was obviously presented as an object of fictitious improvisation, an embellishment of the human portrait in action. The android, as well, offered a unique insight into the impotence of human imagination since, during low cycles in our evolutionary progress, the only effective way to look at ourselves in the mirror was to create a mechanical mirror image. Invariably, the robot entered our consciousness at times when we were, as a species, trapped in our own self-loathing and contemplating the future course of our civilization. The android came as a subconscious teacher and was able to contrast the vastness of evolutionary possibilities by projecting our deepest levels of consciousness onto the screens of abstract hallucination.

Most impressive of all, the closeness between a technologically advanced android and a human being, it turns out, is much smaller than ever anticipated. Humans can be both biological and technological, as long as the level of technology is outside of conventional science. In other words, a technological human, android, cannot exist within the confines of convention and tradition; rather it must necessarily exist under the impositions of fictional science. What is fictional science? It is the kind of science that we know will exist one day, even if not today. For example, the handheld communicator device employed by Captain James T. Kirk in the 1960s TV series *Star Trek* was an invention from fictional science, but an even more advanced touch-screen version, Apple iPhone, is widely used around the world in the modern day.

It is fairly straightforward to imagine a new kind of advanced robot on film or a novel, and quite often is the case that the writer concocts a dystopian future whereby robots are a threat to humankind and must be battled against in a race to save the world from a technological takeover. But really this is a result, likely, of the internal psychological conflict within the artist who cannot reconcile the distance between an unknown, but perceivable, future and a chaotic, and ugly, present. The artist then infuses their current parameter of fear and extrapolates that into a dark vision of the future, and all of it is well within the expectations of a society that lives with a profound and pervasive need of fear-based entertainment. It is expected that any robot of the future is a robot of terror and enslavement. This was the basic storyline of James Cameron's *The Terminator* film series whereby Armageddon was the inevitable flip of a technological switch and the activation of the self-aware Skynet computer system. In order to prevent a human uprising against the machines, a cyborg, embodied by former bodybuilding champion Arnold Schwarzenegger, is sent from the future to the time of the leader of the resistance before he was born. By eliminating his birth, the machines would achieve total human dominion.

Without question, the impressive tales on television and in movie theatres are not as impressive as the reality facing us on US prime time television, only that the obviousness of certain people, chiefly in politics, has been able to prevent their true identities from ever being detected. As early as the 1950s, the creator of *The Twilight Zone* TV series, Rod Serling, was able to integrate man and machine into one cohesive archetype, the living android. Serling demonstrated in a number of fantastical episodes (eg *The Lateness of the Hour, I Sing the Body Electric*), that, with a slight reworking of the emotional range, a person of flesh and blood could have been the result of genetic and mechanical engineering. In fact, Serling probably invented the cyborg (Season 4, Episode 103: *In His Image*) well before Cameron ever left the trucking business to become a fulltime filmmaker. In one telling scene from *In His Image*, the main character, suspecting that something is wrong with him after surviving a car accident, opens up his wrist and discovers mechanical parts and wires under the artificial flesh, a scene nearly identical to Cameron's cyborg 20-years later which opens up his cybernetic wrist after a car accident. And Serling followed closely, if not exactly, in the fictionalized footsteps of Isaac Asimov who had his famous short story collection, *I, Robot*, published in 1950, nearly a decade before *The Twilight Zone* first aired.

While it can be prophesized that, either in print or digital format, an artificial robot, of any size or distinction, will be involved in some destructive or anti-destructive behavior as a result of some trigger event, such is not the case of the androids I discovered on Capitol Hill. Whether they were created in an underground installation using synthetic DNA sequences or the product of advanced robotics commonly found onboard an interstellar class starship, these "Capitol Hill androids" may be hard to fathom and difficult to process, but without exception we are dealing with an entirely new kind of existential culture that, although is within the human spectrum, it is at the same time well outside of our indigenous robotics and genetic theories and practice.

As an inept scientist who has never been imprinted with what is acceptable scientific practice and hypothesis, and having a lifelong fascination with my imagination and intuition, it seemed natural that it would take this kind of pedestrian anomaly to make such a historical leap. Perhaps it is the case that scientists and roboticists alike have been defeated at the creation of robots "in our image" simply because they necessarily rely upon codes of conduct and established scientific principles; for had I any of this conventional education I would not have accepted my own discovery with any sense of validation. My scientific ignorance enabled my observations of the living androids and therefore I might have to conclude that no properly trained scientist could have ever made this discovery, and never has, which is more proof to my point.

This is both a good thing and not a good thing. It is good because it tells us that scientists have been fairly rigorous with their methods of experimentation and analysis.

Similarly, this is not good because it goes without saying that they may have missed much more than the androids on Capitol Hill, and we don't know at this point how much has been missed. As much as this could seem as a negative, what is fundamentally true is that society probably has been given more than it can handle and had rogue scientists revealed all of their tricks we might have seen a societal collapse. It could be for good reason that some of Tesla's ideas never made it into the public sphere, things like wireless energy delivery systems. Imagine if terrorists hijacked an electromagnetic emitter and zapped into oblivion a few undesirable nations. So the context of the situation always needs to be maintained despite personal preferences. It goes without saying that my android discovery is merely a fraction of the advanced technologies already in use in all aspects of our daily lives. It would be naive to think that synthetic humans came about in scientific exclusion, rather what is more likely to be the case is that there are complementary technologies that we will slowly be able to uncover, for example, nonlocal communications systems that can program and deprogram the androids.

Writers, then, are wrong when they suggest that robots are from the future and that they are a threat to a biological society. Certainly this is not my view on the matter. The presence of androids in modern day on American television proves, quite remarkably, that the future is now; that there is a discrepancy between the world we see and the real world. These androids are perception adjusters who can teach us to modernize unlike ever before, and since they haven't entered society with guns blazing we can gain better insight as to the intent of their makers. What is ultimately true is that they have been, at least in my current sample, inserted into the highest levels of power in the most powerful nation in the world. This cannot be an accident by any stretch of the imagination although the low esteem human might want to diminish the significance of this observation.

We are presented with a cultural situation unlike ever before, one that contains both biological and technological elements. While standard science has prohibited us from ever discussing living androids as a real possibility, it would seem natural to at some point have a genuine shift in the scientific community and to examine, without prejudice or fear, these and other examples of artificial intelligence. Of course, this will all have to take place within the context of not only a chaotic world, but, more importantly, within the understanding that the makers of these androids have remained undetectable for good reason, and probably do not want to be discovered, *for good reason.* And so anyone who wishes to find more answers will have to be kind of individual who has the durability and stamina to see things through no matter what the obstacle or circumstance, for otherwise, as my own life has indicated with scars as reminders, the quest is doomed to fail.

The following android investigation is rooted on the material observations commonly found on online videos, prime time television broadcasts, and special announcements and presidential statements, including, but not limited to, political emergencies. It was important to build a theory and even a set of conclusions based, not on proprietary and classified materials, but on public materials with good public access. Both television and the internet are widely distributed enough to justify this requirement, and those without these things easily enough can gain access through an extended family network. Plus, these video materials can be copied and disseminated for a relatively low cost, either through email and mobile phones or through DVD and other stored media distribution. All of this is necessary because without at least a peripheral view of the evidence it is hard for most people to come to proper terms with this kind of phenomenon. And with public materials already available, it is now nearly impossible to stop the dissemination of knowledge, at least in its current form and quantity, and this will eventually lead to new theories and hypotheses within more scholastic groups and societies. The public nature of this discovery is extremely powerful and we should take every advantage of it if we are to get to the kinds of answers that have eluded humanity since its own origins.

ANDROIDS ON CAPITOL HILL

Joe Lieberman   Nancy Pelosi   Max Baucus

The public materials that contain the primary evidence of
the androids, two of them recorded via video camera from
CNN broadcasts on prime time television, contain a rich
assortment of the chief elements that will make up this
investigation. This small collection of evidence has
provided the basis of this discussion with some added
materials detailing a third android subject, the female
robot ("fembot"). It is important to be reminded that
these three robots, forming the basis of this entire
discussion thus far, are not military models as one would
have seen in a blockbuster film, and probably also why
they have remained innocuous. These androids are
political models, and there are other models not

discussed, that offer the best observational evidence I can think of. If we fail to understand this set of evidence, it is nearly impossible to discuss the other cases that are to be found, and widely available to a trained observer. This *android theory* is vitally important to the human race in that it explains by its intrinsic nature exactly what human beings are made of, no matter the race, which undoubtedly point to a new approach to the genesis of mankind. It also takes the fiction out of science fiction and places the story at the dinner table, so to speak.

Screen image (wide) of "Androids Among Us", October 8, 2012 (YouTube).

Primarily then, this fringe medical investigation is based on a set of recorded observations on television, crime scene evidence, and those observations have been decoded to some extent in order to foster a better understanding of what has been existing in front of us without detection. The earliest observations were publicly discussed in four (4) shorter documentary-style videos—the "White-Haired Android" series—posted on YouTube in August 2010.

Those videos led to the release of the larger format 41-minute documentary, *Androids Among Us* (authored under my former name Talessian El-Wikosian) on December 26, 2010 (Boxing Day).

Screen image (zoom) of "Androids Among Us", October 8, 2012 (YouTube).

The follow up public chatter led to an unpolished 104-minute *Discourse on Persons Artificial* on January 15, 2011. It can be said that Talessian El-Wikosian, even though he is still a version of me, is the man who (technically) discovered the androids on Capitol Hill and my theories and discussions follow his pioneering achievement. Many thanks to Talessian El-Wikosian.

I feel it necessary to explain the metaphysical consequence of this android extravaganza, knowing full well that a number of eyes are going to roll. Metaphysical discourses are not things I've shied away from and, at this point, I'm not going to start to reduce all of my experiences into conventional propaganda modalities. The man who rightfully discovered the androids on TV is Talessian and by all indications Talessian is gone, but there are good reasons he is gone and, as well, reasons why I'm not (even though we are the same people); because by right any person who crosses paths with the most powerful agency on the planet, and aggravates them enough, isn't going to live a long and healthy life. Let's face it, this knowledge is more top secret than extraterrestrial ships and technologies, which they are willing to kill activists that raise too much awareness. As soon as I released my first full documentary to the public on Boxing Day 2010, I was allocated for assassination. I know that because I was told that. By whom? By multidimensional friends of mine.

Talessian El-Wikosian had irritated all the wrong people and the consequence was death. I, being Talessian at the time, know I would be killed several months after I posted *Androids Among Us* on YouTube. They give you those several months so that people don't see any immediate connection between your death and your discovery. It's an accident and not an assassination. "He had a heart attack in his sleep. It happens." With Talessian marked for death, and without any conventional way to avoid it, I kind of had to decide how to continue living. If I myself killed Talessian, without physically killing him, before their programmed assassination took effect then their mark of death would be effectively nullified. How to kill someone and still live on? Change your identity. By assuming a new identity, in line with everything else in my life, I would not only get a new life, but; better, I would continue to live, so that's what happened.

There was a price paid in full for this revelation, on many levels and of many kinds, and I've paid it because, if for no other reason, I'm the only one who could afford it. It may be the case that I have made errors here and there in the analysis, but they are minor, and we won't understand them anyway until we upgrade our sciences; regardless, I take none of this information lightly and I take offense to every person who refutes it without at first examining it. The cost to America's children could very well be the end of their liberty and the continuation of a lifetime of economic, political, military, and cultural enslavement. The agencies who are able to accomplish what I only describe in brief in this book are not the kind of agencies that make mistakes. They have long term goals that are met one way or another and they can foresee all the bumps up ahead. It is they who invented the term "pre-emptive strike" because like all magicians it is they who do not like it when the brightness of the future disagrees with the darkness of their forecasts. I urge every American to overlook the trivialities and wake up to the fact that your nation has been woefully taken over by minions of a very powerful cosmic command. America, are you listening? America... your nation has been hijacked by androids.

There will be those who expect a clean investigation with logical deductions and rational observations, and this is entirely normal, but this is not a normal discovery and, in fact, could never have been explained using conventional scientific principles, although in my defence I have added as much as my intelligence and education allowed. As much as this discovery is well outside my education and competence, unfortunately the educated and the competent failed to ever make this kind of discovery; therefore, without my elusive observational qualities there would be no discussion, and there never has been an android discussion of this nature. To this extent, we have to be willing to modulate some of our expectations, myself included, in order to appreciate this kind of rare phenomenon. If Watson and Crick were afraid of serpents it is likely that they would have never discovered the DNA molecule. If businessman Ray Kroc didn't like hamburgers he wouldn't have franchised Richard and Maurice McDonald's hamburger stand into the world's largest hamburger restaurant chain.

Much of the discussion here is based on the *Persons Artificial* 104-minute video presentation and all of its related research. This was updated and revised 9-months later in early October 2011 and given a more appropriate title, *American Androids*, to better reflect the main thematic elements of the material—these three artificial persons were after all American.

(Due to copyright issues with some of the material, these videos may or may not be available for any prolonged period of time unless they were copied and re-uploaded by other users. *Androids Among Us* and *American Androids* (edited version) were also released in DVD and video download formats.)

The original book *The Android Conspiracy* also saw its title change to *American Androids: Special Edition* in order to better reflect the nature of the situation. The *American Androids* documentary was released in a 3-part (160-minutes) video series online in October 2011. I had decided to preface the documentary with a statement of clarification in order to regulate the intent and focus of this kind of irregular work. With the American tag, it came to my attention that this investigation could be misinterpreted as a purposeful attack on the American culture, which it isn't, it just so happened that the evidence I accidentally fell upon took place in Canada's neighboring country, and it was also the case that the nation the androids infiltrated happened to be America. I think this preface bears repeating here:

> The author of this work wishes to state that he is not against the United States of America and has nothing against the three key persons identified herein. This groundbreaking discovery is primarily about an interdimensional group of scientists who are using advanced techniques to

hijack America and key aspects of the world. The key intention is to expose their android agenda.

Video Evidence

Because we are using a more static medium, print, it is important that I describe the primary contents of three key video recordings (observations) that will form the basis of my 'android theory'. Following this phenomenal discovery I have collected, as a natural course of interest, quite a number of other video samples. Although these may be referenced indirectly, to sharpen aspects of the discussion, they are not included in this book, since this book is centered on the initial observations starting in 2008 and my level of understanding has since improved. The reader, if interested, will find more advanced discussions on the subject in the Commentary section later in this book.

One of the most challenging aspects of this work has been the allocation of focus and attention, two things that I have been trained to operate in a more conventional fashion since youth. The natural conventional leaning of my focus meant that I had to pay particular attention to the discovery, and specifically to these unprecedented observations. The focus on the original video evidence, once again, is just another part of the principles I incorporated into my investigation, among other principles and protocols. Again, some of this stringency might come across as arrogant or assumptive, but, given the distractions of the world, this is something I felt necessary in order to preserve the integrity of the crime scene.

Image capture from Video #1. Joe Lieberman (left rear flank), Max Baucus (right rear flank). Conference highlighted the successful approval by Congress for the release of the TARP funds. 2008

The first video, *Post-bailout Press Conference*, presents the most complete observation, especially in regards to the blinking phenomenon, chiefly the **White-Haired Blinking Man**, of the walking and talking human robots. The identification of the politician is Senator of Connecticut Joe Lieberman (born February 24, 1942), a Democrat with "more than 20 years in the United States Senate" as his website indicates. Parent to four children and twelve grandchildren, Lieberman is also the Chairman of the Homeland Security and Governmental

Affairs Committee and a member of the Senate Armed Services Committee, a powerful and influential Senate committee with legislative oversight of the Department of Defense and nuclear energy security. At the time of this writing, Connecticut Senator Lieberman is serving his fourth and last term, and plans to retire from politics in 2012.

The second video, *Healthcare Conference*, provides a very telling account of the second individual, the **Large-Eyed Man**. At the time of the recording, President Obama was pushing through his Healthcare reforms. This person was not easy to identify since I am not familiar with American politicians per se, and have no interest in politics as a general rule unless there are key global issues under discussion.

Image capture from Video #2. Max Baucus. 2009

It was many months later that I was able to identify this person, made particularly difficult because he appears to shape-shift from video to video and I wanted to be sure that the person in the video matched the person in the official portrait, as Montana Senator Max Baucus (born December 11, 1941 as Max Sieben Enke). He is a Democrat, the fifth longest-serving US Senator in office, who is the chairman of the Senate Committee on Finance and Joint Committee on Taxation, and a member of other committees and subcommittees.

Image capture from Video #3. Nancy Pelosi.

Finally, the series of clips titled Video #3, *State of the Union (2007-2011)*, provide samples of the **Blinking Woman**. This is a remarkable woman in many more ways than is fit for discussion. At the time she was the Speaker of the United States House of Representatives, the highest-ranking female politician (a Democrat) in American history and the woman first contacted one Thursday night in September 2008 by Ben Bernanke, the Chairman of the Federal Reserve, who said to her, regarding an economic emergency, point blank: "We won't have an economy by Monday." Bernanke was referring to the looming economic fallout that would eventually lead to a rush-to-the-pump financial rescue plan. Nancy Patricia D'Alesandro Pelosi, now the Minority Leader of the House, was born on March 26, 1940 and has five adult children. Pelosi blocked the impeachment proceedings against President George W. Bush stating that impeaching the president, having misled Congress about the weapons of mass destruction in Iraq and for illegal wiretaps on the American people, was "off the table."

The *State of the Union* address is an annual report made by the elected President to the United States to Congress and the people of the Republic. The Blinking Woman (Pelosi), the 60th Speaker of the House, is found in consistent form in the *State of the Union* addresses during her years in office (2007-2011) spanning from the Bush to Obama Administrations. She is consistently blinking, along with the occasional eternal stare (and emotional cheer), in these high-profile televised speeches starting as Madame Speaker, to roaring applause, on January 23, 2007, alongside Vice President Dick Cheney. Even in 2007, a one blink per second is noticeable to those who have demonstrated a little patience, but this is only the beginning of the Pelosi tale.

Only Video #3 was recorded from the internet on YouTube.com. Videos #1 and Video #2 were recorded from live TV in my Vancouver, British Columbia bachelor suite from late 2008 to early 2009. Length times are approximated and do not include the full conference, speech or presentation, as I did not record the entire presentation. The State of the Union addresses are typically 60-minutes in length, some are as short as 50-minutes (2007) up to nearly 70-minutes (2010). Collectively, averaging one hour per show, the State of the Union speeches contain 5-hours of video evidence on Pelosi alone, a character central to my android investigation, and significantly erode the objections that her anomalous mannerisms are isolated cases that are dismissible as normal behaviors. It is fair to say that certain aberrant behavior should be dismissed, and even overlooked, as regular aspects of a person. But when a person, on prime time television and in front of an audience of the highest-ranking people in American power, not to mention the tens of millions of live viewers, exhibits anomalous ticks and odd movements we have every interest to pay closer attention, and rightly so.

Video #1: Post-bailout Press Conference (11:00 Mins.)

Video #2: Healthcare Conference (5:40 Mins.)

Video #3: State of the Union (5 hours)

Video #1: Post-bailout Press Conference

An Elderly Politician (unidentified) is speaking at a press conference on Capitol Hill, Washington, DC. The elderly politician is flanked by two other politicians creating a triangular formation. One on the *left rear flank* and one on the *right rear flank* of the speaker. The politician on the right rear is the only one wearing glasses. They stand in a V formation.

ELDERLY POLITICIAN: My preference would have been Senator McCain's Bill of $450 billion dollar of tax cuts... in the legislature you don't get everything you want... but this bill has a very substantial component of tax cuts... the Republican moderates were able to see to it that more than a $100 billion was cut from this program... now there are people who would like to spend less, some would like to spend nothing... I believe that the position of the United States Congress is a solid position from a very conservative organization that a Republican group very concerned with fiscal restraint...

I turned on my Kodak Z650 camera and clicked on the movie-mode function. I knew that it would be hard to get a clear picture on a television set (due to flicker rate) but the picture picked up the scene quite well. I wanted to record it so that I could study it later. It was an intuitive response, there was something else going on, rather than a logical one, having no other information presented at the time. The first thing I noticed was the white-haired man on the *left rear flank*. He has a large forehead, is slightly balding and is over 65 years of age. He looks like a typical US politician, like a man who has been on Capitol Hill his whole life, his two piece suit and tie are picture perfect. The press conference, on the usual level of observation, looks and appears as completely normal. The politicians are making their usual promises, tossing figures in the billions like rice at a wedding and they all seem congratulatory, itself an oddity on Capitol Hill. But the white-haired man is blinking in a very rapid, rhythmic manner, a manner that for all intents and purposes appears to be mechanical as I focus my inexpensive camera, capable of capturing 640 x 480 video, on his face to see just how peculiar this blinking is (Also read *Appendix: An Unscientific Blinking Test, p.229*).

The post-bailout press conference was recorded sometime in October 2008, I am fairly certain of that. It could have been early November, but I know it followed the September 24 prime time address of President George W. Bush and it followed the euphoria of President Bush approving the fiscal stimulus proposed by then-Treasury Henry Paulson.

The entire Wall Street bailout was odd, it was pushed through with a supernatural haste, all the parties seemed to agree with one another and if not, by the next day they had come to an agreement. President Bush signed the Bill/Act into law only a few hours after it passed Congress. The first proposal by Treasury Paulson was rejected. That created a concern among his friends and they frantically reworked the financial proposal, a very long document that couldn't have been properly interpreted, and it passed on October 1, 2008. The public nearly revolted the idea of bailing out the failing, and corrupt, banks with their hard-earned tax dollars and with no promise of a return. This prompted the monotone speech, a well-crafted if not simplistic conversion text, by President Bush as a means of convincing the masses that this hasty, improvised, and highly irregular situation needed an immediate, do-or-die decision as soon as possible.

GEORGE W. BUSH: ... This rescue effort is not aimed at preserving any individual company or industry; it is aimed at preserving America's overall economy. It will help American consumers and businesses get credit to meet their daily needs and create jobs, and it will help send a signal to markets around the world that America's financial system is back on track. I know many Americans have questions tonight. How did we reach this point in our economy? How will the solution I propose work? And what does this mean for your financial future? These are good questions and they deserve clear answers...

The general atmosphere of the 2008 Wall Street Bailout was extremely tense, hasty, and imposing, there was a decision to bailout the banks and that idea had to be forced across the board and the public had to accept it however it was presented, with the condition that their plan would work and that America would be saved. History tells us that very little of their plan worked as promised and, more so, there were so many subplots to the general rescue story that to this day no one is certain what really happened. In Pelosi's mind, and likely in the minds of Bernanke and Paulson, the two sorcerers behind the ensuing financial chaos, and retroactively excused of any liability or responsibility, the economy was diverted away from one possible collapse into one real collapse: double-digit unemployment, more than 42 months of recession, housing foreclosures unheard of, and a never-before drop in the credit rating. All of this happened under President Obama's watch. Obama would later run for re-election in 2012 and blame the economic crisis on his predecessor, forgetting that he fully endorsed the foolproof Wall Street bailout (and that figures from Goldman Sachs, an investment bank that received funds, also backed his presidency).

The politicians continued the threat that without a fiscal stimulus, the economy would collapse, unemployment would reach record figures, and, from my recollections, it had the feeling of the end of the world. The economists who rejected the bailout plan were not listened to or left on the bench. It was the end of America without the bailout. That was the overall theme, that was the overall message, and they kept playing the same doomsday song over and over again.

This recording following that theatrical stage play: "We are going to bailout our banking friends, we like the idea and everyone has to agree, if you do not agree, the economy will collapse, you will lose your house, you will lose your job and we will all suffer, what do you think?" It was typical of the Bush Administration to push policies through, the same George W. Bush who, along with his friends, pushed through the need for a military invasion in 2002-03. Saddam had weapons of mass destruction and he needed to be stopped or else America would suffer terribly from biological weapons, or possibly improvised nukes (dirty bombs). It was the end of America without the invasion. Colin Powell, then-Secretary of State, artfully made a presentation out of thin air to the UN Security Council on February 5, 2003 to produce valid reasons to invade Iraq, claims that were later discredited by the Central Intelligence Agency (CIA).

Not long after the invasion took place, it was discovered that Saddam Hussein had no weapons of mass destruction (WMD), no mobile biological labs, and that the Bush Administration had purposely fabricated evidence (made everything up) in order to go to war against Saddam's regime. Top aide Powell regarded that presentation to the UN as a low point of his life. From a CNN interview with Powell published August 23, 2005, "I look back on it, and I still say it was the lowest point in my life."

The Wall Street Bailout had the same tinge of purposeful and malevolent manipulation. These high-ranking players were in a rush, there was an air of urgency, and they intended to get whatever they wanted using whatever words that would induce the necessary approval, as a doctor who would use the hormone prostaglandin to induce labor. And they, like a doctor with an overdue pregnancy, got what they wanted.

I was lucky that finance and banking, my least favorite subjects in university, were still not my particular interests, for had they been I would have been using my calculator to see how much this bailout would cost American citizens. Instead, I was already on the fringe of thought, I was developing my basis for *reality science*, a new kind of physics that would determine the architecture of a manufactured reality. I was involved in discussing interstellar cultures, what I called 'Stelans', and had spent considerable effort working on my own UFO Disclosure process, which wasn't as much of a disclosure as it was an attempt to clear up as much as the disinformation as I could handle.

On a side note, there was good reason that late 2008 would turn into an opportunity to have an official UFO Disclosure from the new US Administration. Incoming President Barack Obama seemed to be an ideal interstellar candidate to lead a disclosure process. Bush had revealed himself to be a warmonger and a puppet, and clearly had challenges with sobriety and the ability to communicate as a general function of his leadership (he was prone to gaffes which provided prime material for late-night comedians).

With him in power, the truth of the presence of Stelans would never come out; he wrapped the minds of citizens with war and terror. Then on the eve of a transition in November 2008, a bit more than a month before Obama became president there was this 'crisis', an end of the world, a do-or-die decision. Obama, a Democrat, and Bush, a Republican, happily agreed on the Wall Street bailout, without so much as a single conditional sentence. In fact, as it magically turned out, everyone agreed. Our *blinking white-haired man*, Lieberman, who appears to be the highest in authority, spoke after 3 speakers during the press conference:

JOE LIEBERMAN: Thanks Senator Reid... working together under Harry Reid's patient, usually soft, but occasionally quite tough leadership, we came a long way in a relatively short time to achieve something big and urgently necessary for our country and our people, and when I say "we" I say the President, the House, the Senate, members of both political parties — *everybody* gave something in these negotiations to achieve something bigger for our country and our people. I personally am confident that this Economic Recovery and Investment Act will put a floor under this economy of ours, it will be the beginning of the turnaround for the American economy, it will protect, or create, millions of jobs, it will put billions of dollars in the pockets of everyday hardworking American people so that they can go out and spend it and grow the economy. This, in my opinion, is a turning point, we're not gonna get back to where our economy was overnight, but I think this represents the beginning of turning our economy around and leading to a better day full of opportunity for the American people. I was proud to work with this group, proud to be part of it, without the three republican colleagues, the Senate could not have come to the table... without a great spirit of compromise from the White House, the House and the Senate we would not be able to announce this agreement today.

Image capture of Video #1, Senator Lieberman.

Though the white-haired man is identified as Senator Joe Lieberman, for my exotic independent investigation, his individual identity is not important. His position is important. He is a high-ranking Senator in the US government. I have no argument regarding the US government, nor do I carry any ill will about Senator Lieberman and my personal observations of him, and my discernment of his artificial characteristics, could in fact be inaccurate. In fact, it will be somewhat inaccurate until we reach a point whereby we can verify in some meaningful way the things I have discussed.

As well, I have no financial interest in the Wall Street Bailout; but, as we will see, the bailout plays on another level of manipulation. As much as this book relies on three American politicians and even aspects of the US economy per se, in no way is this book against the nation of America. That is neither the intention of the book nor the purpose of the author. America is a great nation, just like every other nation. She has the advantage of having a higher level of development and a higher world status in this period of the human civilization. But the only real evidence I have are these three individuals and they are positioned at the highest levels of the US government. In this regard, and in the light of my developing android theory, this must be presented and discussed. Had it been another nation, I would have done the same thing.

I had avoided the initial three synthetic specimens for a number of, what I thought, valid reasons, and if anyone was to reference the original book editions they'd see the adoption of anonymity and to focus on the actions and characteristics of the androids. As strong as these feelings were during the critical period of the discovery, after moving months and years away, and understanding the necessity of identification in the viewer's mind, I have decided to more completely identify the androids. This comes with a conditional statement: my investigation is specifically tailored to the android characteristics of these individuals, since I do not know them, or care, as individuals, and certainly have nothing against them. I should also reiterate that I am a Canadian citizen. As such, I am just as concerned about the anomalous presence on Capitol Hill as should each and every American citizen. It is likely only the result of a lucky draw of cards that I happened to be the first man on the scene, and I would think that the reverse would also be true had an American detected a possible threat in my home country.

In the abridged version of the transcript, notice three key 'elements' in the speech of Sen. Lieberman—President, House, Senate—all of which came to an agreement, a unanimous handshake from a very diverse set of people who, under normal conditions, do not agree, and in fact like to disagree, argue, and make accusatory remarks to one another. In fact, they get paid to argue and they work hard to represent a diverse constituency. Likewise, it was well known that the public was quite angry over the Wall Street bailout and their political representatives, had they been keen on their constituents, would not have bent over so readily. Luckily, for my unnatural investigation, their compliance helped significantly to put the pieces of the puzzle together that much faster. There wasn't much gray here except for the hairs on people's heads.

We should also take note of these talking points because they will come into play later on: 1) turnaround for the economy; 2) create millions of jobs; 3) billions of dollars in the pockets of hardworking people; and 4) a compromise of White House, House and Senate. This is an amazing achievement and an even more amazing promise. We've heard this rhetoric many times before, but there is an additional dimension this time, there is the presence of artificial persons.

Senator Joe Lieberman is blinking rapidly in a rhythmic movement for the entire 11-minute clip. It should be noted in the World Book of Records, "Man Blinks for Eleven Minutes While Holding a Press Conference." He at least could earn an award for blinking at a rate of two blinks per second for most of those eleven minutes.

Alongside Lieberman are four speakers. An elderly politician speaks first, he is followed by a female politician then she is followed by Max Baucus at the *right rear flank*, the Montana Senator we will discuss on Video #2. Lieberman is the fourth to speak. When he speaks his blinking becomes less rhythmic, irregular, almost normal, and yet if you study him closely you can see that the blinking conflicts with his speech pattern. He is still blinking unusually fast but not at the high-rate during his position in the left rear flank. Regardless of the interpretation, the visuals are rather impressive in and of themselves, indicating that there is more than one stream of information processing. A fifth speaker, a Senator, speaks after Lieberman.

The oddities of this clip are many actually. At first I noticed the rapid blinking of the white-haired Lieberman. Then I noticed, as the clip plays on, Lieberman is positioned in the left rear flank of the speaker. There are four speakers besides himself (some are off-camera). As each speaker speaks, Joe maintains an almost disciplined stance at the left rear flank, and aside from an occasional facial expression or comment to a colleague, he is in strategic position and is blinking in a rhythmic fashion. The blinking is not irregular; it doesn't have a natural characteristic. In fact, it reminded me of REM (rapid eye movement) sleep stages, when the eyes experience rapid movement while a person enters a short period of deep sleep characterized, unlike non-REM sleep, by brain activity equivalent to those when a person is awake. Only that Lieberman is fully awake, during his REM sleep, and in the middle of a very important (and televised) press conference, following a madly-rushed bank bailout; and is blinking excitedly for nearly eleven minutes straight.

In addition to Lieberman, the large-eyed Baucus on the right rear flank, wearing glasses on this clip, is maintaining a pleasantly still pose, almost statuesque. Looking at his eyes, he appears to be blinking slower than normal, as if he is not blinking enough, but not staring either. There is a speaker, with what appears to be normal blinking, and there is Lieberman on the left rear flank blinking and, on the right rear flank, there is Baucus who is not blinking, with a masked face. Should this all be considered normal behavior? In the minds of the millions of TV viewers and the mass of colleagues, it is surprising that none of what I have so far identified stands out, and has never been worth mentioning in any significant manner. Among the community of intelligence and federal agents, their primary function to scan the population and its representatives for potential threats, including identifying threats through psychometric observation. This duty is of particular interest to our investigation because this kind of supernatural observation normally falls under the jurisdiction of national defense. But it could be the case that the intelligence community was too preoccupied with wiretapping the civilian population, all potential threats to the nation, that they missed the blinking Lieberman on national television. As well, we do not know how many of these anomalies have been missed and for how long they have been going on. If Lieberman's 20-plus years in politics has given him the position with which to supranaturally influence American policy, we can infer that he and his controllers have taken every opportunity to alter the course of America in favor of some mysterious agenda.

Video #2: Healthcare Conference

The non-blinking Senator Baucus in Video #1 is also on the clip in Video #2. He is not wearing glasses this time. He is positioned on the same right rear flank of the speaker as in the previous example. The speaker this time is President Barack Obama, a far more significant player on the political field. The large-eyed statuesque Baucus is virtually unmoving and unflinching throughout the entire 5-minute clip. He is unblinking. He wears a masklike appearance; he could pass for a statue or a man suffering from Parkinson 's disease. The recording is shot Live on CNN, same as Video #1, and my recording is as it is being broadcast on TV.

Image capture of Video #2, Senator Baucus.

As President Obama continues his healthcare promotional speech, what will later be pushed as Obamacare, the glaring Baucus stands in the same position without fail, with an occasional slowed blink. I focus on Mr. Obama's eyes as he speaks; he appears to have a *regular* amount of blinking, we can say 25-30 blinks per minute. The stark contrast between the blinking speed of Mr. Obama and the frozen expression on Baucus' face is evident in this clip. And because it is 'live', there is no mistaking the evidence. It is what it is. Baucus, flanking the powerful man in the world, is acting in an extremely uncanny, outlandish manner. He might have blinked several times a minute.

If Mr. Obama blinks 30 times a minute, Baucus may have blinked three or four times, and all of them falling at the tail end of the minute. In other words, Baucus doesn't blink for 50 or 55 seconds and then gives four shallow and fast blinks. This pattern seems to repeat itself. The nature of the presentation is also important, in contrast to the previous bailout, because in this circumstance it is Mr. Obama who is pushing his healthcare reforms and perhaps that is why an android is employed so as to assist in this challenging matter. Obamacare is an important policy that is pursued throughout Obama's first term as president, later to be passed in a new bill, and no wonder the androids were employed to get the job done. Why use a human to do an android's job?

Since we know that the financial rescue didn't rescue any of the citizenry, and we know that androids were involved in the plan, and since we have noted another android in the healthcare reforms, we can assume that the healthcare reforms under the Obama Administration are not in the best interests of the people of the United States. Even we don't exactly know all the reasons why the new healthcare system is a bad idea, we simply know that these illicit and illegal androids are involved, and by default their plans are not in the best interests of regular human beings. We saw this in the bailout and we can see this in the healthcare reforms.

Furthermore, no other political pushes can more appropriately justify my hypothesis, which is extremely important to our cause, in regards to the deployment of artificial politicians: androids are utilized to alter human behavior and opinion in order to reprogram society to accept a scenario that, under normal conditions, they wouldn't normally accept. When there is a tough policy to convince taxpayers of then get out the androids, charge their batteries, and start reprogramming the sheep.

Video #3: State of the Union (2007-2011)

To contrast Lieberman and Baucus, and to reinforce my investigative analysis, I decided to include Nancy Pelosi, as she annually sits on the left rear flank of the president during the State of the Union addresses.

High angle on Pelosi during a State of the Union.

Looking at several samples of her on various State of the Union speeches, including those under Bush, years earlier, and Obama, her excessive blinking is clearly evident. Her blinking is well known even in comedy circles like David Letterman's *Late Show* whereby Letterman counts the **number of blinks** between her and the non-blinking then-Vice President Cheney over a 30-second video clip. Pelosi's rate jumps into the low thirties while Cheney barely moves a facial muscle.

Pelosi on CNN's *The Situation Room*, October 2012.

Again, I have nothing against her and I do not know her, but she is blinking in a rapid fashion, about half the rate of Lieberman. We have Lieberman clocked at a rate of two blinks per second, Pelosi at one blink per second, and Baucus at no blinks per second. As well, Pelosi is positioned on the left rear flank, same as the heavy blinker Lieberman. What is it about the left rear flank of an important US political speaker? These are the State of the Union speeches with President Obama or President Bush doing the honours. The audience includes members of Homeland Security, the Central Intelligence Agency (CIA), the Federal Bureau of Investigation (FBI), the Pentagon, politicians and other important dignitary and military figures. The audience includes the major decision makers in the United States. Of course, I am less concerned about what President Obama is saying, or his audience, than I am concerned about the rapid blinking computer terminal situated behind him—a woman who is one of the most powerful women in America sitting on the left rear flank of the most powerful person in the world in the capital of the most powerful nation on the planet. If she is an android, as I have determined, then this is a matter of national security.

These three (3) videos will form the basis of my makeshift medical investigation. I think they hold enough evidence to push my 'Android Theory' forward. In investigating the presence of persons artificial, I decided to adopt a medical and scientific analytical process on a foundation of science fiction possibility. The science fiction approach is visionary, but typically incomplete and being limited to the logic of the storyline, providing only what the protagonist and antagonist need.

In stark contrast, my observations of the three androids on Capitol Hill posed an impossible set of intellectual gaps, mostly because all great thinkers and the entire planet believes that life is entirely predicated on biology.

What I witnessed first-hand, and recorded on camera, was evidence of an advanced mechanical quality allocated to certain human beings. These people did not fit the regular human groups and, as we will see, they didn't fit the mentally unfit humans, the institutionalized people who suffer from neurological tics, spasms and hallucinations. So, where did they fit? And where they best fit was in a new class of humans, artificial persons, synthetically manufactured people who were perfectly disguised as humans. So, what were three artificial persons doing in the government of the most powerful nation in the world?

CRIME OF THE MILLENNIUM

The 2008 Wall Street bailout failed to deliver on every promise. By 2010, the American economy was in the midst of a *suppressed* second Great Depression, unemployment hit the double digits, homes were foreclosed or abandoned, but the banks, the automobile manufacturers, the politicians—they were all fine. In my investigation, I traced back the pattern of the 2009 Wall Street rescue to the year 1929, around October 29. Along the way, I noticed some other surprising characteristics. As the android hypothesis began to take shape, during the process of looking at neurological conditions, and as my video observations confirmed my artificial suspicions (and unscientific hunches), it became clearer that the financial stimulus to save America was probably better interpreted as a way to commit a large robbery without a trace because everything the financial rescue was designed to do, proposed by the financial mastermind Henry Paulson, Secretary of the Treasury, and his banking consultants, in fact it did the opposite. And sitting in the midst of this criminal master plan were, as I will argue, three artificial persons who appeared to be completely human.

What is more impressive than robbing a nation than if you could rob a nation using synthetic puppets? This is the Crime of the Millennium, isn't it? And not the first of its kind. These synthetic puppets were in the middle of an economic crime. The proof is that the results from the master plan two years later were the complete opposite of the promises made. I'll go into this later on. The impressive part is the fact that I have identified three political persons, all of whom were involved in the financial stimulus fiasco, all of whom were salespeople for the master rescue plan, all of whom still have their jobs, and, more importantly, all of whom have been *artificially constructed.* That contribution to the 2008 fiasco, these artificial fingerprints, is the advanced DNA evidence I needed to crack the case of the millennium.

The Wall Street Bailout then became a crime scene, not just any ordinary crime scene, an extraordinary crime scene that used advanced technology against a comparatively primitive society trapped in the limitations of a 3D box. These three people were involved in the hijacking of a nation. The most powerful nation in the world. Though the crime is important, this is a *fringe medical investigation.* If I can prove that these three people are androids, I will not only have proven that the world you think you knew is not the real world, I will also have shown you how they committed advanced crimes without anyone ever knowing. Additionally, I will have allowed us to realize a hidden dimension never before seen.

If these three people are androids, as I in fact stated in my hand-made documentaries, then that means that there is a mysterious culture hidden among us, a culture that is significantly more advanced than us. A culture with the science and motivation to build artificial human beings. It is without a doubt that Lieberman, Baucus, and Pelosi all appear to be real persons, what are referred to as 'human beings'. And what I am going to do is to explain why these three persons, although seemingly, perfectly normal, although appearing to be biological and intelligent, are in fact *persons artificial*, androids of a very advanced quality. They are machines of a very impressive quality, made from a quantum technology far ahead of any human science. And they are being employed to hijack society without society ever noticing a thing. The most perfect crime in human history. Our crime scene is Capitol Hill.

If you wanted to invade a nation, without spilling a drop of blood and without shooting a single bullet, employing a group of programmable androids is an extremely efficient method with which to achieve your fanatic goals. The geopolitical framework of the world kind of lends itself to takeover. The United Nations has 193 members. Why is this figure important? The United Nations represents 98.5% of the world. In a way, the world is centrally-planned as if out of some kind of improvised communistic model: central banks, central security councils, central religions, and central space agencies.

The world kind of lends itself to an easy takeover, or, the world has already been taken over and these centralized facilities have been positioned as a way to keep the rest of the world in order. The presence of three high-ranking political droids in the most powerful and influential nation in the world, as we shall see, is not the result of some evolutionary accident. Darwin's theories are no longer valid in this discussion, and, frankly, Darwin's ideas died in 1940, the year that Nancy Pelosi was born because androids are not born like normal people, androids are created, they are manufactured, and that means Darwinism has been dead for over 70 years, just no one found out till now.

The handmade documentary *Androids Among Us* was a frank discussion, even if shorter than expected, on my discovery on androids living among humanity incognitos, best of all it was free of charge. I had concluded from my observations that on those recordings, those three identified persons were synthetically made, and that it was a level of technology that was way over our heads. I estimated at least 1,000 years ahead of modern science. This designation is nonlinear because scientific discovery can make leaps and bounds over previous ideas, but it is far enough ahead of human science that we are still short a few giant leaps for mankind. And it was so advanced that it 'appeared' to be biological. We will see this theme a lot throughout this investigation; things appear one way in this context and appear another way in another context. For example, a doctor could view my video evidence and make some kind of autoimmune diagnosis.

A conspiracy theorist might watch the blinking Pelosi behind the president and start to think that it is the end of the world.

Video #1 and Video #2 were recorded from a television screen as it happened, on live TV. Video #3 was recorded from the internet, some YouTube channel. There are no special effects. I did all the recordings myself and moved the camera around to isolate the oddities I was noticing. From the earlier chapter, you get a taste of the context of the conference—everyone is coming together to save the country, everything is going to turnaround; money will flow into the pockets of the people. It won't be an instant reversal but there is an imminent reversal, America will be saved.

The alternative was made very clear: if the bailout is not approved, the nation might collapse, it will, and might never recover and you might lose everything you own. This is a very serious situation. There are 311 million lives involved. There is a great nation on the chopping block. There is a perfect rescue plan. Oddly, the *only* rescue plan and it is *guaranteed*. If the plan is approved, if the banks and the corporations are bailed out financially using the people's tax dollars, the United States will be saved. Similarly, if the bailout is rejected, the American economy will collapse and a second Depression will take place, and America will never recover. It will be the end of this great nation. If it was a psychological operation to force people into making difficult decisions using the gun-at-the-head approach to leadership, under the militarily-inclined

Bush, then the operation worked. Force people into a corner and they will fall to their knees and throw you their wallets. If only Jesus had been so lucky he wouldn't have had to be nailed to a wooden cross.

Much of the following discussion will involve Lieberman, Baucus, and Pelosi. These are only my best examples at this time, perhaps my most convincing evidence at that moment. I have no prejudice against the US government. Neither do I have any judgement of any of these people I refer to. Their individual identities are not in question. I do not know them personally. It is what they are made of, the *manufacturer* if you will, that is in question. Because there are many more androids in the world.

Further, although all three examples are politicians, android persons are not only politicians. These political examples highlight some hidden aspects of a most interesting game. Because political figures are influencing society, influencing leadership, there is investment aspect to this game, an agenda, and the key to that agenda are LBS—three persons who are working for the most powerful nation in the world. Whoever created these persons, these elegant and intelligent puppets, and they truly are works of art from master craftsmen, has an investment in the direction of the American nation. It wouldn't be the first time. Let us next focus on the artificial mechanics that is hiding underneath the biological shroud. Perhaps I can earn some of your trust in this most auspicious investigation.

FRINGE MEDICAL INVESTIGATION

This is not the kind of investigation we would find in real life, though it is possible that we might find it a blockbuster film, because androids, and living robots for that matter, have yet to exist in the modern world, but roboticists are hard at work in trying to resolve that technological gap. We are only at the beginning of the investigation with many puzzle pieces still not put onto the board. Without someone who is willing to undertake these kinds of imaginative thought processes, we might never see any measurable movement toward the establishment of true artificial intelligence. My investigation is going to be a kind of imaginative medical investigation, even a fringe medical detective story.

In my fringe medical investigation, I have enough evidence within these three subjects to arrive at a number of conclusive remarks. This is not your usual evidence. I am going to stretch our understanding of life. I have to because otherwise there is no other way to explain androids on modern day television, and this is what I have decided I have seen. This is not what I wanted to see.

I just wanted to watch TV. So this is what I have seen and this is what I have come to understand. We cannot move forward into an undetermined future without moving forward with untested approaches to undetermined objects.

Here is my key medical evidence:

1) The rapidly blinking (and positioning) Joe Lieberman
2) The catatonic mask and stare of Max Baucus
3) The long-term rapidly blinking Nancy Pelosi

Their collective traits:

1) Over 70 years of age
2) Senior-level US politicians
3) All in favour of the 2008 Wall Street Bailout
4) All closely connected to the US President
5) In favor of a new healthcare reform bill ("Obamacare")
6) Somewhat aware of their artificial qualities

This is a fringe medical investigation and I am going to play the chief investigator for things that are not supposed to exist. There is a lot of supporting evidence as well. What happened was that these people made some kind of realization; they realized that their dominion over people was wearing thin; or, more importantly would be compromised. They needed a counter move in order to defend their terrain. It is possible that a counter agency got involved, that they made some kind of play, or intended to make a certain geopolitical move, and that triggered the phone call from Bernanke. It could not be the case that the Chairman of the Federal Reserve and the Treasury Secretary, and all of their executives and servants, that all of them did not foresee the imminent economic crash so well described. It would be illogical to think that. At the same time, there are elements we do not know, and cannot seem to know in any detail. We cannot know what triggered the economic crisis, only we do know that it did not happen the way it was explained to us. It could not have happened that way because that would mean that the economic masters are woefully incompetent. Under normal conditions, this is what we would naturally think, but for some reason we have lost our rational thinking and have decided to believe the story as described, even though it doesn't really make sense.

I certainly cannot accept that all the economic masters in America did not foresee the immediacy of their financial needs. Therefore, it makes sense to think that there was a mysterious trigger that we have not been made aware of yet. That trigger forced the sorcerers to make a play, an improvised financial rescue plan to disguise their true intentions. Their true intentions could have been a play to buy the support of some external agency or society that would be able to negotiate with the unknown trigger or threat. I am speculating, but if this were the case, in order to afford the services of this other agency the economic masters needed to pony up a significant amount of funds, within one or two weeks. That is what prompted the emergency phone call. That is what prompted all the political parties into widespread agreement, since it is likely the case that all of Congress prays to one sort of master, and then it only seemed natural that their chief puppet made contact with the people. The advantage of this supernatural imperativeness, and it does involve supernatural elements, provided me with the kind of advantage that is rarely ever seen, likely never seen. They got sloppy. Instead of following their regimented protocols to hiding among us, they loosened some of those protocols, and needed to, and that was when I just happened to notice, and I happened to record it, and many months later, I happened to have figured it out.

On the surface, as it might appear to most people, these are rational, logical and normal speeches. The politicians are doing their job; they may or may not be aware of what exactly they are doing. They may fully believe that they are going to rescue the nation. They may think they can buy time and then rescue the nation. There are likely a handful of players, as usual, who are rigging the deal. But as my perceptions remove the biological shroud, these persons were merely puppets to a very intelligent and hidden hand, an entirely new group I refer to as an advanced scientific race of people; a race that is likely more advanced than extraterrestrials; a race that has never ever been discovered in the history of this planet.

There is no trace or record of the android manufacturers. There is no record or mention, ever, of androids on the planet, though we can infer through rigorous study and examination that there have been strangers present in society that could be considered to have synthetic traits. For example, the inspiration behind Mary Shelley's 1818 novel *Frankenstein: or, The Modern Prometheus* could very well include the manufacture of synthetic people, only done so using reductionist thinking. Shelley may have seen something synthetic in her experiences, visions, or dreams, and aspects of this may have come through her creative writing. She wouldn't likely have been able to fully comprehend androids; instead, she translated her inspiration into something that she could understand, the stitching of dead (synthetic) parts into a new human. This is not unlike the stitching of DNA sequences to create the ideal genome. Shelley may have witnessed genetic engineering, and having no real capacity to understand its molecular details, her neurological wiring provided the closest translation it could find. The result was the compilation of corpse parts into a newly-formed man, child. What reminds us about the impressiveness of this story is the enduring qualities of this 200-year-old story.

Even today there is no android factory or city that we have been able to identify and expose. And yet, on live TV in 2007 and 2008 and 2009 and 2010 and 2011, there are living biological androids, and they are over 70 years of age. That means they have been around since World War II, and no one noticed. No one. So, it makes perfect sense that most of us are blind to this technology. That we might need a chance to attenuate ourselves to what a science 1,000 years ahead of ours might look like. At the rate of scientific progress, with synthetic cells and artificial intelligence, it isn't far into the human future where a living android could indeed be manufactured. At the current rate of technological progression, we could have a machine being in 20 years and an android in 50 years, assuming we have no unexpected technological leaps or advances.

What I am presenting isn't as far away as it appears. The problem is our view of life is biologically-based and limited to the 3D range. These androids on Capitol Hill appear to be biological when they are not. This will dip our discussion into human origins as well, for if the human being can be replicated to a level that is indistinguishable from a "real" human being then with only the right science and carbon materials you could create an Adam and an Eve.

None of this makes sense. It won't make sense if we restrict ourselves to the 3D box. So you will hear me say often that we have to step out of the box in order to make sense of something that is outside our range of understanding, of something that has never happened before. As I have stated, this is also a crime scene. What's the crime? Androids are being used to change the course of a nation. Not just any nation—the most powerful nation in the world. They are essentially HIJACKING SOCIETY. Instead of hijacking an aircraft and flying it into a World Trade Center, they are hijacking a government and driving it into a Depression. A government that is carrying 311 million passengers.

And they are using artificial people, androids, ladies and gentlemen, simulacra who look and act human. Androids that, until I pointed them out 'were human'. Actually they are not human, and were never human. They were created in a lab, they were conceived in a test tube, and they were birthed and inserted into our world. Because they look like us, and because we haven't learned to identify their nuances, we have adopted them into our various cultures. Androids are not only Caucasian; they come in every race imaginable, it's just a matter of eugenics. The clues to how they remote control these androids will take us inside the brain. But before we get into our neurological view, let us first review some tactical methods the androids employ.

ANDROID TACTICS

There are two specific androids in the Post-bailout Press Conference (see image above). Lieberman is standing on the LEFT REAR flank of the speaker (unidentified). Baucus is standing on the RIGHT REAR flank of the speaker (all in relation to the speaker). The trio form a V-shape, a kind of triangulation, and throughout the press conference they both maintain those positions, as wingmen, unless they are speaking. Notice their positioning. A triangular formation. The white-haired man on the left flank is *blinking in a rapid state*. On the opposite side, the android is holding a *catatonic glare*, an *unblinking* state if you will. This is very clearly evident in the image and it is even more striking in the video sample. But does this fall under normal human behaviour?

Why is it when Lieberman on the left, positioned on the left flank of the speaker, an action he maintains throughout the press conference—and I have only recorded a portion of it here—that he experiences a rhythmic rapid eye movement? This white-haired politician is wide awake and in the middle of a nationally-televised press conference, a distinction that more often than not is overlooked. The context of this phenomenon is just as important as the phenomenon itself, that for a career politician of more than 20 years civil service to behave carelessly at a time when the US economy is on the verge of collapse is ridiculous. This is not an exaggerated error, and neither is it the result of a medical condition, as we will discuss later in this book. What has been captured on screen is a purposeful manipulation known well in android culture.

While the speaker is speaking, the positioned Lieberman is blinking frantically at a rate of 2 blinks a second. On the recording, it is obvious that his rate of blinking is exceptional. It is also evident that he blinks rapidly when he is in the rear left flank position. We also notice that whenever he has an emotional response, a gesture with his colleague Baucus, or a thought, his blinking is interrupted; otherwise, he blinks almost supernaturally. You can imagine a person in REM sleep with his eyes wide open, only that the movement of the eyes during sleep is irregular and Lieberman's eyes are particularly rhythmic. When he moves to speak, the rapid blinking stops, replaced by a mixed variation of blinks, still fast just not as fast as before. When he speaks, his blinking is irregular, closer to normal. But as soon as he plops into position once again, his eyes almost immediately drop into a state of rapid eye movements.

Video capture from Video #1.

We contrast the white-haired Lieberman on the left with the large-eyed Baucus on the right (see captured video image above). He is *not* blinking. He wears a masklike face, even more explicit in Video #2. A facial expression that lacks any real response. A catatonic stare. He could easily be mistaken for a statue. But when it is his turn to speak, Baucus is eloquent, loud, poised, his beautiful voice shines through. Had he been suffering from catatonia, or some neurological disorder, he wouldn't so easily be able to shift into political speech. He probably wouldn't be invited to speak at a televised event had he been dealing with a recurring illness. Instead, he is attending the event and he is speaking. He is also maintaining a masklike expression, similar to someone with Parkinsonism. Does he have Parkinson's? Ex-boxer Muhammad Ali, diagnosed with Parkinson's syndrome in 1984 (in his early 40s), wears a similar masklike expression on his face with one key difference to Baucus: Ali cannot transition into a normal speaking voice with normal facial expressions, unlike Baucus who can seemingly shape-shift on live TV without so much as a hiccup. Of course, this is how it appears and ultimately we won't know exactly all the details. Watching Baucus seamlessly transition from mask to man appears completely normal and undetectable, as it should, as it was designed, but when you understand the true nature of a synthetic heritage, the transition becomes an impressive display of artificial science. His genetic designer is a master, not only for him, but for all three of my roboteers.

Again, I am not saying Baucus, Lieberman, or Pelosi have any autoimmune disease. They may or they may not. Baucus may be on medication, as millions of people are, but it would not be information that I would have access to. Further, the three of them may not even be artificially prepared though this is the basis of my speculative hypothesis. What is evident in the recording is that Baucus is statuesque for most of the recording and yet speaks without pause. If it is an illness, he is in full control of it; and any person in full control of their illness is either heavily medicated, or the illness probably isn't an illness.

Video capture from Video #2.

In Video #2, as we noted in the opening, the large-eyed Baucus, in splendid synthetic form, spends the entire nearly 6-minute clip, recorded from live TV, not only in a very pronounced catatonic stare, but he is also situated on the same right rear flank as in Video #1. This time there is no person on the left rear flank so we cannot see anyone blinking. What is interesting is that what he was doing on the right rear flank of Video #1 during the bailout conference, he is also doing in Video #2 during the healthcare speech from president Obama.

Video capture from Video #3.

Then we turn to the blinking woman in Video #3. Now, remember how we noticed the white-haired blinker in Video #1? He was consistently on the left rear flank of the speakers. Well, the blinking Pelosi also happens to be situated on the left rear flank of the speaker *and* she is rapidly blinking. Is it possible that both the blinking Lieberman and the blinking Pelosi are by chance sharing the same uncanny eye movements? Is there some uncanny connection by a person on the left rear flank and rapidly blinking? The evidence strongly suggests that this cannot be a coincidence. In the medical field, people may share similar symptoms and may even experience similar side effects from the same medicine. But when it comes to standing positions, the likelihood of Lieberman and Pelosi sharing saccadic movements is extremely low. The chance of an unrelated man and woman having the same nervous twitch or the same disease while in the same tactical position, 'the left rear flank' of an important political speaker, is extremely improbable.

During all five of the State of the Union addresses selected for this discussion, the leader of the Nation (starting with George W. Bush in 2007) is flanked by the Nancy Pelosi. Not just any woman but a sharp dressed human-looking and high-ranking politician in her own right. A robot that may or may not know it is a robot. If my analysis is correct, there is good reason to think that one or more of these robots are aware that they are robots. The word 'robot' need never be used, and is probably considered a derogatory term; I use it occasionally to highlight the synthetic base of these creations. A robot may be colloquially referred to as 'special' or 'different' or simply 'not like everyone else'. In fact, it would be in the interest of the android makers to instill in the specimen some categorical term so as to ensure that the robot never diminishes itself into humanistic thinking and therefore remains a powerful conduit for behavioral control.

These active robots, sitting at the highest levels of political power in the most powerful nation in the world, are specifically inserted in order to treat and adjust human behavior; certainly they are not there for window dressing. The robots look human by design so that they can more easily interface with the indigenous population, people who would never believe in human-looking robots. This impressive set up is engineered to be a self-defeating exercise. Advanced artificial intelligence pods are inserted into a society that has been trained to disbelieve in advanced artificial intelligence pods. Should any rogue member of society ever wake up and attempt to inform society of the robots at large, society, brainwashed to disbelieve such fictional ideas, will only ostracize, and in the old days crucify, the messenger.

In a strange twist of fate, with a tinge of Stockholm Syndrome, it is the slave who is defending their master from their own emancipation. The master needn't make any comment because they have conditioned the slave population what is real and what is false; and the slaves, by simply acting upon those false programmed ideas, deny themselves their own escape. Only a very advanced mind could concoct this order of illusory enslavement, which is more justification for the presence of truly advanced thinkers that have yet to reveal themselves. And even though they are still largely concealed, their fingerprints are all over Lieberman, Baucus, and Pelosi. On any other awakened planet, this is enough evidence to win the case in favor of humanity. On this planet, this is a provocative and surreal discussion that is usually best suited for the world of fiction. As far as I can tell, these three politicians don't just contain the fingerprints of the gods but they are the fingerprints of the gods.

It may also be the case that any of these three robots is fully aware of their robot status, even having full awareness that they were manufactured in a scientific lab. At this point I am not certain as to how much these robots know. I suppose because these three were purposely utilized to redirect America that they are probably high-level models. Furthermore, because all three of them are of the same generation, and are all seniors by definition, they have had more engagements than any of the newer models. If I was an android maker and I was going to desperately influence society or risk losing control of my particular nations, I would likely deploy my best bots on the ground. We could infer that Lieberman, Baucus, and Pelosi are top robots for human administration. And since Pelosi is the eldest of the three, and as a woman the most powerful in America, none of which is easily achieved, it could be said that Pelosi is not only a robot, more importantly, Pelosi is one of the master controllers. That would be my best guess.

If my puppet state was under threat, I wouldn't leave it to a remote-controlled robot to do the job. I would want to be on the ground, as the robot, protecting my territory. At some point when the farm is under its greatest threat, the farmer pulls out the shotgun and goes to task. Pelosi was the first contacted and the first warned of the impending financial doomsday, she led the charge, she pushed Paulson's bill through, she told Bush to get involved or else, and he did, everyone from the President to the House and the Senate all fell in line with Pelosi's magic. I can think of no better reason than it was due to the fact that Nancy Pelosi is a disguise for one of the master controllers, for only a master could exact such a comprehensive control of an entire governing body in such a short period of time. The average person, unaware of their enslavement, will attribute her success as a result of good men and women who wanted to rescue America and to protect the homeland. The truth, however, is clearly more difficult to discuss. As a more general observation, theoretically the other two robots also have some awareness of who they are. Specifically, I would identify Baucus as to having a moderate ability to recognize his own powers, and Lieberman, less so. I think it comes down to processors and memory systems. Baucus has an amazing ability to shape-shift on TV and this cannot be executed without some form of mastery and mastery requires conscious effort; therefore, I think Baucus is perfect for the task because, like Pelosi though not as much, he too realizes that he has magical properties, whether he hears voices in his head or just seems to know what to do and say, this man is in close communication with the control center, and is in awareness. In this sense he has an advantage over the sleepwalking population.

Lieberman appears to be a more mundane model, a robot that is designed to lower human vibration, and therefore a man who is himself unable to reach high levels of awareness. His task appears to be hypnotic, putting people to sleep so as to keep them under control whereas Pelosi is a firecracker, she can light up the entire White House given the chance. She is a powerhouse that even when wrong, she is right—so much so —that she is always right. I would attribute this to her high energy nature. She emotes arrogance, likely a reflection of her artificial vibration. A Pratt & Whitney turbojet engine idles much higher than a Toyota automobile engine.

The top US military brass, Secret Service, CIA, FBI, NSA, scientists, homeland security experts, advisers, consultants, and audience members—none of these have figured out that the anomalous person behind the podium is a fully-functioning android. Why? Because she herself hasn't explained her artificiality and that most of the given audience has been indoctrinated in a system which by default denies the existence of living robots in the modern day. Plus, she doesn't look artificial. She's a beautiful mature woman robot. By the end of this book, hopefully, you will begin to think that she and the other two gentlemen are in fact a new class of humans, persons artificial.

Again, I have nothing against any of these people nor do I care, or not care, about the exact operations of the US government. What they do, in the normal sense of the word, the duties of all civil servants, is primarily their business. What concerns me, and why I have taken it upon myself to investigate this phenomenon, is that there are some serious incongruities with the Wall Street bailout and because I've now decided that there was some advanced AI interfaces that were deployed to deceive, and those interfaces were so advanced as to have been completely undetectable by any stretch of the human imagination. That meant that an advanced agency was involved and I felt obligated, as a moral duty, to assist human awareness because the best and the wisest hadn't noticed anything strange other than perhaps another political coup or public deception. This was more than a political coup, this involved androids made with a technological hand that was at least 1,000 years ahead of humankind and that meant there was a scientific group, still unidentified, that had the ability to create living, breathing, procreating androids. If only Philip K. Dick was still around to see his hallucinatory dream states actually coming to life.

I often get asked about this 1,000 year attribution. How did I come up with this figure? The figure came about by compiling human technological achievement over the last two thousand years and then comparing those accomplishments with those of the last 100 years, and then taking that comparative analysis and contrasting it with the technological achievements of the last 25 years, using the introduction of the internet as a nodal reference. So you have a new technological messiah in the name of Jesus who is then used to formulate an entirely new religion called Christianity some 1,700 years ago. Christianity is not just a religion but it is a scientific approach to life, citing ideas like resurrection, heaven, and eternal existence. These are scientific ideas as well as spiritual ideas. The next technological leap came when Tesla and Einstein, and centrally distributed electricity, came into our lives. These were powerful leaps from Newton and Galileo thinking models. I compressed the

Christian waveform and Einstein waveform and the internet waveform together, taking all of their data along such as computerization, space exploration, and thermonuclear detonation, and I then measured that up against living, breathing robots, even using films as futuristic references to produce my approximated technological future of 1,000 years. I don't think that humanity could handle a 1,000 year jump in technology at any one time. That itself might lead to massive chaos or collapse. For example, if you were to have handed a nuclear warhead to Napoleon Bonaparte, a French leader from the 1800s, he'd very likely have detonated it in 1812 during the French invasion of Russia, and have altered the course of history. But several technological leaps spread out across several decades that would be enough to achieve the necessary principles in android science. Additionally, a 1,000 year technological leap is not equivalent to a 1,000 year time jump since I view technological progress on a nonlinear timeline and time as a linear measurement.

Think of the internet, first invented by Tim Berners-Lee in the late eighties. The early adopters had no graphical browser, no video downloads, no wireless networks, and certainly no online shopping. Look at the adoption rate of the internet as a communication medium and the hundreds of millions of handset devices, smartphones, all equipped with wireless interconnectivity. From the HyperText Markup Language (HTML) of the eighties to the multi-touch screen, 6th-generation iPhone 5 (released September 21, 2012) and its 700,000 available apps, we have endured a technological leap and are none worse for wear. The best mobile phone in 1983 was the Motorola DynaTAC, a 750-gram brick-sized phone with 30 minutes of talk time and a price tag of $4,000. The DynaTAC had no internet, no video, and no music player. It was just a phone. The iPhone 5, on the other hand, weighs 112-grams and can do everything except cook your dinner and pour you a cup of coffee, but give it time and it might be able to do that and a lot more.

I mean, if there are three androids on primetime TV then *someone* made them, and if they are over 70 years old then someone made them *70 years ago*. That means that this millennial science was available long before the mobile phone and the laptop computer. You can see reasons for my obligation. I felt obligated to explain the crime of the millennium. I cannot control the result except to insist, and to explain, that the androids are not the prime interest, the prime interest is their maker. I want to know who created the Lieberman, Baucus, Pelosi android models. Did they create some of the human race as well? Do they have the cure for all disease? Do they know the meaning of life? How often do they talk with God? You can see how important this is. I have just discovered an iPad tablet device in the hands of Mary Magdalene, and I am the apostle Peter trying to tell people that the device in Mary's hands isn't a metal cutting board. And Mary is being uncooperative. She doesn't want people to know that she has a tablet computer, and Jesus is certainly going to defend Mary, his alleged wife, and his best disciple.

Personally, I don't care all that much that hardworking Americans were getting fleeced. It's been happening for many decades, around the world. The war in Iraq was a monumental abuse of power and none of the criminals were ever charged. It seems that Americans don't mind being fleeced as long as they can have beer in the fridge and weekends off. Their freedoms seem to be dwindling by the hour and their greatest complaint is that the economy needs to be fixed. The economy was broken by the very people who are being asked to fix it. It's sort of like asking the thief who stole your car to help you find the thief who stole your car, for a retrieval fee. They are the same person. Bruce Wayne and Batman is the same person, but they are not going to tell you that because that would compromise their dual lives.

The androids are not going to reveal to anyone that they are androids, without question their makers will never speak a word about it. That is why I have taken it upon myself to help them take off their masks, and to prevent any more fleecing of the great American nation. I like all nations. I like all races of people. I happened to have discovered androids in America, lucky they are close neighbors to my home country of Canada, and I, as any concerned global citizen, am presenting my best ideas in order to preserve the sovereignty of the United States because I think that the presence of these three androids, and the clandestine ease in which they were deployed, indicate that America has already been hijacked by an extremely advanced technocracy, the likes of which is more likely found in a Dick dystopian novel.

I care that advanced technology is being used to shape society, a society with global influence. I care that an advanced race of people is hiding behind the walls of reality and I don't know if they believe in Jesus Christ. Maybe they manufactured Jesus Christ too? Maybe they were there 2,000 years ago and they taught the meek to build wooden crosses. If they are 1,000 years ahead of modern science, by my estimation, then 2,000 years isn't all that far away, is it? How far back do they go? They likely have time-travel technologies as well. They might be from our future, or from our forgotten past. These people may be from a place so foreign to us that we may never understand because those memories may have been erased from our collective unconscious. We might never know who they are because they removed their existence from our brains and perhaps that is why they have remained undetectable. Perhaps their detection is rooted in our ability to recall our lost memories of them.

Why the crime became important is because the crime presented us with a pattern of behaviour and that pattern had a historical link. That gave these advanced races of people a historical connection to human evolution and progress. If this link held, that proved that these advanced beings had an investment in the direction of this civilization. How long have they been instructing mankind? Were they behind the wars? Were they responsible for the pandemics? Were they there for the fall of Rome? What do they want from the primitive human race? How many of their androids are here? Is there a ratio of androids-to-humans among the population? Before we continue, I'd like to revise the definition of an android to this:

> **android** *n.* a living organism built with synthetic DNA appearing as a normal human being (having any racial characteristic).

The android thinks it is alive because it has been hooked into some electromagnetic matrix and therefore it thinks it is alive, as a computer thinks it is alive when it is connected to the internet. We are going to have to superimpose our understanding of androids with a science fiction model of awareness. Why we need to do this is because we have been conditioned to believe that biological life forms are alive and that artificial life forms cannot be alive.

We cannot say "dead" because experiencing death is only attributed to life forms that were at first alive. To our current thinking, an android cannot be dead. It cannot be dead because it cannot exist. My position here today is to allow you to perceive that those individuals in my video recordings from primetime TV, whom I have selected as my primary examples and offered to the world court as primary evidence, that those individuals are androids—that they are in fact *artificial*. That is the first point. The second point is that they are 'alive'. But we have to understand what kind of artificiality they are made of.

We don't naturally believe in artificial life. We believe in biological life. Artificial things move (eg automobiles) and think (eg microprocessors) but they are not alive. The technology in your mobile phone is more advanced than a mainframe computer of the sixties and yet your phone still needs you to make a call. The cellular phone cannot think and talk unless it is pre-programmed to do so. It is perfectly natural for us to rely on 'biology' and to put aside 'nonbiology'. But in this case, I have living proof of three artificial life forms on TV and they are not only talking, but they are thinking and breathing. They are well-dressed, they have jobs, they have friends, they eat and drink, they sleep, they have children, and they are artificial.

In order to bridge those biological gaps, I will briefly turn to a few stories. Hopefully, this will give us some possibilities, even if fantastical in nature. It isn't important to be completely real. Why? Well, because this discussion has never taken place in the history of this planet. We have no precedent and that means we are forced to step out of the box, way out of the box.

STORIES OF ANDROIDS

To the viewer, the android appears alive, it has a voice, it makes decisions, it breaks down, it grows—for all intents and purposes it is alive. And the job of any good puppeteer is convincing the audience that their puppets, their marionettes are alive. They need you to believe their puppets are alive because they need to tell a story, they need to convey information, to share an allegory, to teach. And the puppet is a good medium for communication. As the Canadian pioneer Marshall McLuhan famously said: "The medium is the message." The adoption of androids as modes of communications, of which we have so far only discussed in very broad strokes, especially given the restrictions of the evidence thus far, is perfectly in line with the engagement protocols of advanced nonhuman societies. In fact, it might be a form of interaction, in terms of first contact directives, that is the safest method of evolving primitive cultures. A primitive culture, in this context, is any culture that has not yet mastered artificial intelligence, because we can assume that any race of beings that can manufacture robots are very likely themselves robots.

We are fundamentally not going to find any people of the flesh in going forward as a civilization. Given the impressive qualities of these three examples, and their human likenesses, added to that the fact that these robot models were designed to fit into a primitive civilization, we can only imagine what kinds of body types are available for truly advanced civilizations. We are fortunate to have had a number of visionary storytellers share their science fiction stories with us. We owe a great gratitude, at least I certainly do, for otherwise we would have no references with which to relate to what has been hidden among us.

The first great story that tackles the life of a puppet is found in Florence, Italy, in the year 1881. It is the story written by Carlo Collodi about a wooden puppet named Pinocchio. Woodcarver Geppetto carved him from a piece of pine wood. The book was *The Adventures of Pinocchio*. Pinocchio is a wooden puppet who gets into all manner of situations and who desires to be "a real boy." It's not so much the story of the puppet that is important, rather it is that the puppet has dreams of being real. There is this exploration of a synthetic puppet that can talk and move, and who dreams of becoming real. In this case, becoming real means becoming mortal. As a puppet he cannot die since he is not even alive per se, but he doesn't think of his immortality; instead, he dreams of a mortal life. Is Collodi referencing the incarnation of a soul into a mortal body?

He is essentially portraying Pinocchio as this happy immortal soul who wishes to incarnate into the flesh. Could this be a dream about his own incarnation? Do happy and immortal souls dream of becoming human, so much so that when it is granted it is a wish come true? And, if so, then why are some people so miserable that they were born on this earth? Have they forgotten that it was once their dream to come here in the first place?

About 5 years later in France, the author Auguste Villiers de L'Isle-Adam writes a novel, *Tomorrow's Eve*, that features an artificial human-like robot named Hadaly. According to an online quote, we get the glimpse of some new information, information that will help us later on. An officer in the story says, "In this age of Realien advancement, who knows what goes on in the mind of those responsible for these mechanical dolls." It is an interesting quote because it presents some crucial ideas. These ideas are relevant to our story here. First, the word "Realien." Apart from the obvious "alien" word dealing with offworld cultures, together with "Re" you also get the conjunction "real" and "alien." And in French, alien comes from aliené, a French word meaning *insane*. Or in Latin, *alienatus*, to be estranged. And a person who studied insanity was an *alieniste*. Coincidently, psychologists and psychiatrists used to be called *alienists*. Then the quote goes on to talk about "those responsible for these mechanical dolls." Obviously, there were people who built the human-like robots. *"Who knows what goes on in the mind of those responsible..."*

This single quote seems to be implying that either the makers have an agenda, a secret agenda, or, it could mean that the minds of the makers could have been transferred onto the mechanical dolls (ie telepresence). That the consciousness of the makers is *imprinted* on the created beings. Imprinted where? De L'Isle-Adam is writing 130 years ago about mechanical people, in the plural. *Tomorrow's Eve* would later serve as the thematic backbone to a successful Japanese anime, *Ghost in the Shell.*

Nearly 100 years later, in 1968, Philip K. Dick published *Do Androids Dream of Sheep?* The novel explored what it is to be human in the future. In 1982, Dick's novel was translated into a feature film starring Harrison Ford and directed by Ridley Scott called *Blade Runner.* Director Scott would later be responsible for one of the most famous commercials, the Apple Computers ad from 1984. It was an iconic portrayal of an oppressed society, a tyrant on the screen and a giant hammer smashing tyranny and then the logo for an Apple Macintosh computer. The tyrant was supposedly IBM and the effect was unmistakably relevant. Apple Computers, with Steve Jobs at the helm, went on to be the rebel with a cause. Apple's latest iPhone incarnations include an intelligent personal assistant, Siri, who helps you with your tasks just by asking her questions. Siri is by all definitions the genie in Apple's bottle. Siri also represents the introduction of a robot in the making. There is a mechanical body, a microchip, circuitry, a network, a name, intelligence, and a voice. In the future, Siri will have two arms and two legs and have the beauty of a supermodel. In 2011, Dick's brilliance was brought back to life in his final book *The Exegesis of Philip K. Dick*, an 8-year documented attempt to explain the operations of the universe. The hypnotic and freewheeling discussions, compiled after his death from his many scraps of notes, in the nearly 1,000-page book, include discussions of machine existence, parallel dimensions, and a programmable reality architecture, closer to some of the ideas found in the fictional film *The Matrix*, only Dick's ideas are based on personal confession and things he had seen in various hallucinogenic states of awareness. Philip K. Dick would have found no fiction in my investigation.

In 1973, the writer of *Jurassic Park*, Michael Crichton wrote and directed a movie called *Westworld*. Starring Yul Brynner as a lifelike robot called Gunslinger, *Westworld* was set in a futuristic amusement park. It was a high-tech, highly realistic, fictional adult amusement park featuring human-looking androids. For $1,000 a day, guests could indulge in fantasy with the androids that were programmed to serve man. For some unknown reason, the androids begin to make mistakes and their programming gets corrupted and the guests on vacation are now trying to stay alive. It is chaos. Kind of like what is happening in our world today.

The 1976 sequel, *Futureworld*, released a poster with this slogan: "Futureworld, where you can't tell the mortals from the machines... even when you look in the mirror!" Crichton was not involved in the project, but it once again is centered on the Delos amusement park. This time Delos is replicating the rich and powerful and programming the clone duplicates, which are so advanced that they are indistinguishable from the real thing, to destroy their originals. The two protagonists (played by Peter Fonda and Blythe Danner) are reporters who are replicated in order to spread good word of Delos through the media.

Essentially, Delos is designed to take over the world. Writers George Schenck and Mayo Simon draw up an impressive high-concept fantasy tale, but fall short in terms of providing an engaging story. I had long forgotten about this movie, I was barely nine years old when it was originally released. I do remember watching *Westworld* on TV a number of years following its release. In those days, films took several years before they were broadcast on national TV.

In *Blade Runner*, Harrison Ford played Deckard, a policeman who specialized in hunting androids, genetically-engineered human beings with a lifespan of only several years. There was a corporation building androids, since they were more durable and expendable, for offworld exploration and sometimes the androids would escape to Earth and disguise themselves as humans. To find an android, corporations would apply an empathy test. To test the empathic response, or lack of empathy. An android lacked the empathy of a human. Ford's character Deckard believed he was a human hunting androids. Even when he returned to the man who made androids, as part of his investigation, he wasn't told that he himself was an android and the earlier versions of the film clouded that issue. But Ridley Scott himself admitted in an interview many years later that Rick Deckard, the leading character, was in fact artificial, but thought he was real. A striking similarity with Pinocchio—an artificial boy who never knew he was artificial and when he found out he wanted to become real. A sequel has been in the works for several years and Scott has indicated that it is a project that may one day see a theatrical run albeit in an entirely new format.

Then we have the Japanese cult classic, a 1995 animated film called *Ghost in the Shell*. In this visionary work, the soul of the artificial person, a cybernetic organism, is referred to as their *ghost*. The lead character is an artificial woman who similar to *Blade Runner*, works for the police. The difference here is that the cybernetic organism is made of both machine and organic components. In *Blade Runner* the androids appear completely human and can only be identified by their empathy. In Cameron's 1984 *The Terminator*, Schwarzenegger plays a cyborg sent from the future to assassinate the mother of the leader of the future resistance. By assassinating the mother, the child who will lead the resistance against the machines will not be born. As the police psychologist commented in the film "sort of a retroactive abortion." Cameron would go on to make *Avatar*, and if the title doesn't give it away, it is a film where alien bodies are genetically-manufactured and the human consciousness is transferred into the blue

Na'vi people in order to survive a harsh alien climate. *Avatar* differs in many ways from its cybernetic predecessors because the beings in *Avatar* are genetically distinct alien life forms, not robots and not androids. This is a remarkable shift for Cameron who cracked open the cyborg flask with his original version of a cybernetic assassin. Several *Avatar* sequels are currently in the works and will assuredly reveal much more of the story.

Let's recap—we have a story of a living puppet, a story of mechanical dolls, a story of an android who thinks he is human, a story of an amusement park filled with programmable service androids, an amusement park that is replacing the influential members of society, a story of a cybernetic organism with a ghost, a time-traveling cyborg assassin, and genetically manufactured people where you can download your consciousness and be reborn. Even on this broad overview you can see that on a fictional level, and spanning across more than 100 years, we have a good premise for the existence of artificial people.

Mechanized people are not new concepts. You could even say that Leonardo da Vinci, and his anatomical drawing of the *Vitruvian Man*, also attempted to blend art, science, and the human body as he determined to illustrate the impressive symmetry of the body. And the study of symmetrical relations between the body parts is the standard procedure for any genetic designer who wishes to clone a particular subject. All of these fictitious elements are going to be useful as we continue this discussion.

FINGERPRINTS OF ADVANCED RACES

I am now going to go back to the video recordings to see how we can explain, in rational terms, the unusual characteristics of our androids. I can't prove how to make androids, at least not without regular access to a secret android factory. I have demonstrated that there are people who are alive today, as a small sample of a larger population, who are probably not human (as defined by conventional medical science); and who are what I think is best described as 'artificial'. And the term android is justified. I did consider adopting new terms such as 'synthetic' and 'artificial' as a better descriptive, but decided to continue to use 'android' since it is clear to me that Lieberman, Baucus, and Pelosi are three of the most congruent androids I have seen, having others to compare them to after the fact, and the android designation highlights their artificial birthrights. Nevertheless, it is not the end descriptive that we will eventually put into usage, but a proper designation that will serve us for the time being.

I do not have the lab to make androids, I have yet to meet the android manufacturers, but I am certain they exist because they have left their fingerprints. Three of them. I have shown you the fingerprints but no one has ever read this kind of fingerprint. Imagine, 500 years ago, I made a presentation on how to read fingerprints or that DNA could be used as evidence in a crime scene. I would be laughed at or excommunicated, depending on the circumstance. But these androids on these video clips represent the fingerprints of their makers. They can tell us who or what made them. The better we understand their technology, the better we can find the makers. To do so I am stepping further and further out of the box. I am hoping we will not lose our way and do something stupid. I am going with you all outside the playground in order to show you that the world you believed was simple is simple because of advanced technology. The more advanced the technology the more obvious the object. The simplest device we can identify, say the leaf on a tree, is the result of an unimaginable technological application, and the fingerprint of a master artisan.

As we will increasingly see, advanced technology appears simple, it is invisible because it is normal, and it is obvious. We all think of it as 'natural'. Those people in those clips appear natural but they are artificial. They are not natural. They are synthetic people and they were made by someone. So, if you agree that they are artificial then you agree that there are some very technologically advanced people that no one knows about.

Not only that but they know how to influence, and hijack, the most powerful nation in the world. And they have been doing it for quite a while, as we shall see. That means that there is a technological hand shaping society: Telling you and your children how to eat, sleep, and think. Convincing you of what is important and what is not. And these technological people are secretive. They are invisible. And as the quote said, we do not know what is in their thoughts.

What is their moral position? What is their spirituality? Do they even have spirituality? What is their agenda for the human race? Why are they so secretive, are they hiding something? Are they monsters controlling artificial human puppets and relaying false stories to society? There are quite a number of questions we ought to have. We've never had these questions because we never knew what their fingerprints looked like. Their fingerprints have two arms, two legs, and multiple bank accounts.

These fingerprints are deciding your tax bill, your defence budget, what your President says, who lives and who dies. These puppets are covering up the truth on aliens and their starships. These puppets are preventing you from seeing them in their full glory. But if I am correct, if these artificial people are the puppets of technological masters, then the world is not run by humans. It's not. It's run by not only a secretive group, more importantly, that secretive group is very advanced. Advanced enough to cloak their presence for centuries, if not thousands of years. That advanced group is actually more advanced than aliens. That their technology could be higher than many extraterrestrial races is a possibility that we've never been allowed to consider. Why do I say that? We have spotted aliens. Millions of witnesses have seen UFOs, and appropriately ignored and discredited by the official decision makers who decide what we are supposed to see. There is a plethora of books on contact with interstellar races of people. Have we ever spotted the programmers? Have we ever realized the androids among us? How many 'nonfiction' books talk about androids on national TV? One. This book.

Where does that put the advanced alien races? We don't know at this point. This is not part of my conversation. What we can see is that there is a new class of android scientists and they are technologically mature enough to simulate life, by creating a line of android cultures, with an amazing precision. These android people are living among us. They are also mixing their genetic heritage with the indigenous population through basic copulation, and, in a way, diluting what remains of the natural human DNA. They are converting man, woman, and child into robot man, robot woman, and robot child. Because of this new narrative on the planet, we have slowly been genetically altered to vibrate in a range of frequency that is within the demands of these master controllers. While we have been inundated with massive distractions, chaos, terror, crises, and every imaginable threat they could think of (well, they haven't yet exhausted all the ideas out of the Book of Revelation), they have been evermore busy doing the one thing that has a far more lasting effect—genetic modification. It isn't only food that they have been modifying, it is humanity itself.

We worry about the GMO foods. We are concerned about the poisons in our water and the push for pointless vaccination programs. But what we have completely missed is the diminishment of our genomic heritage. If there was any great card to play, they've been playing it for at least 70 years (the approximate age of Nancy Pelosi), and likely far longer than that. Seventy years is more than enough to manipulate one or two generations of DNA, depending upon how many androids they inserted during the late 1930s through to early 1940s. I would suspect that during that period, a turbulent period highlighted by evil alliances and a Second World War, a mass of android children (and embryos) were inserted into our plane of existence. Many of us may have DNA sequences in our genome right now that have a predetermined allocation and a native state of enslavement. We may have been preternaturally 'genetically lobotomized', which is truly a difficult thing, not only to accept, but to correct. It is difficult to correct since we haven't a clue as to what DNA sequence is natural and what DNA sequence is artificial. We haven't the instrumentation to analyze DNA at this high level. Plus, they may have inserted inactive DNA sequences into any number of people, through copulation practices or through GMO food, or even through aerosol campaigns and general pesticide use, and when the times were right on the DNA clock those sequences could be activated with either the right frequency or the right food additive. A DNA sequence may be simply activated through a particular vaccination program (mandated by an oppressive regime), buried within the live vaccine formula, or it may be activated through a hidden broadcast signal through a television broadcast, or even something far more incomprehensible. It is nearly impossible to figure out, or to anticipate, what the DNA situation may turn out to be, if any, until we at first accept that there are androids living among us and, in fact, are positioned as our leaders.

Reviewing the medical evidence

There are several logical explanations for the cataleptic stare and the rapid eye movement of each of the three individuals. I'm going to, in order to simplify things, focus on the blinking androids first. What is causing Lieberman and Pelosi to blink?

Here are 5 possible blinking causes:

1) He or she might be a practitioner of transcendental meditation;
2) He or she is a channeller or a medium, people who can channel the souls of people from other dimensions like ghosts;
3) He or she has a medical condition such as a mental illness and he has a side effect from the medication;

4) He or she has an autoimmune disease and naturally exhibit this kind of behavior during key moments of anxiety or stress;

5) He or she is processing dream data in waking state, could be dreaming while awake.

Let's review the particulars: Lieberman blinks when he is positioned in the LEFT REAR FLANK of the speaker. The blinking Pelosi is also on the LEFT REAR FLANK. The type of speaker has no effect on the man's position. Whether the speaker is male or female, important or not important, he remains fixed on that left flank. The speaker, it is noted, is politically important and is trying to influence an audience or change public opinion. Then when they are in that position they experience REM state while awake, eyes wide open. When they go to speak, or have an emotional reaction, they stop blinking wildly. When the man gets into position he starts blinking again. When the woman gets out of her emotional reaction (eg enthusiastic applause) she returns to blinking.

A practitioner of meditation is a very disciplined person. They are usually very calm, loving and intelligent. They usually put aside time for meditation. And if they meditate long enough, if they become a master they might retire early or teach. Lieberman is a long-time political figure, he has never announced his love of transcendental meditation and he would know that a press conference taking only 15 or 30 minutes is not the time to impress the media of his meditation skills. And if this was the case, his bosses would have told him a long time ago not to meditate during a press conference. I think we can rule this one out.

He or she might be a medium. I have seen mediums, I have seen people channel 35,000-year-old Atlantean souls, I have seen some rapid eye blinking for limited periods but that was in conjunction with heavy breathing, burning incense and a quiet audience; a certain ambience. If he or she was a medium, why would they stand behind a political speaker in the middle of an important press conference and conjure up some soul from some other dimension? It doesn't make sense. Plus, their bosses or journalists would comment. Notice that no journalist has ever commented on Lieberman's eye movements as being unusual. Is it normal to experience REM in waking state while on the left flank of a political figure? Sure.[1]

Our third reason why Lieberman is blinking includes a medical issue. He might be on medication that has the side effect of blinking. Some side effects can be quite unusual. It doesn't fully explain how he can experience the side effect while in the left rear flank and not experience the side effect when he talks, but let's explore the medical condition as a possibility because as you will

[1] I have omitted nervousness since I have determined that the rhythmic blinking, quite evident in the clip, is not a natural nervous twitch. From my research, I noted that an irregular movement could indicate a disorder of some sort, including a nervous twitch. His blinking is clearly rhythmic, like a machine. Plus, he is a seasoned politician. I have seen him in other clips, he does not demonstrate any nervousness, he is trained to speak publicly, he has been doing it for many years, he has been on camera thousands of times. I ruled out any nervous related issues.

see, it holds some interesting new information. In my best judgement, both the unusual static face and the machine blinking represent physical proof of some internal neurological workings. These conditions can be observed very readily and I think that it gives much credence for a more thorough medical analysis. There are, admittedly, other less visible aspects to these video recordings, things that we would require machinery such as electromagnetic scanners, to detect. And there are also metaphysical aspects which we can infer based on our medical analysis.

Due to my limited understanding of neurology, I will be forced to keep the technical words at minimum (which perhaps makes this material more digestible). By the same token, my imaginative speculations will not be so limited and I will have (probably) a greater access than usual to an assortment of other reasons for their medical conditions. Again, this is not a usual situation and the only way to understand it is to be unusual.

The rapid eye movements and the mask will now lead us into the brain. I am not a neurologist nor am I conventional scientist.[2] I have observed something quite

---

[2] Properly trained scientists have a much more regimented, disciplined and organized approach. They would do a more thorough research on a number of other similar cases; they would detail those observations and compare observations, etc. But a properly trained scientist would never come to the conclusion that there are androids on TV at the rate of my own discovery. In a way, it is a thing only a pseudo-scientist could achieve. If my observations and conclusions are close to the mark, you can expect properly funded scientists looking into this matter. I have no budget, no medical degree, no orthodox approach, and that might just be why I noticed what no one has ever noticed. The trade-offs seem worthwhile.

   The chief issue with proper scientific research is the time involved in the process. I felt that time was of importance, that to begin the discussion today,

unique and I will use what little understanding of the human brain I can to see if I can extrapolate on how to operate an android. If I can do that then I think my theory becomes more rooted in scientific possibility. Again, I am keeping my own observations and investigation restricted to a very small set of evidence—the blinking and the catatonic stare—primarily because otherwise this investigation might get out of hand.

I want to build a case on the evidence provided or observed as much as possible. There are likely many other things going on. I will comment on some of these and I will offer my remarks. I think if we understand some of the implications of this situation, and the context of the technology, we might have a better perspective. In the meantime, we go to the human brain.

---

even in a more rudimentary form, could have a significant impact on society and that this might more easily allow properly funded research projects further from tomorrow. It was an amazing discovery, in the day and age of a YouTube, I found an immediate need to communicate to my audience. This book is now an extension from a series of video presentations, so I can expect, as long as my hypotheses are fairly accurate, that artificial people will be a growing interest.

BIOLOGICAL BLACK BOX

Whenever we deal with motor skills, we turn to the nervous system, specifically the brain. The 3-pound brain is a very complicated chemical circuit board. Oddly enough, the brain has a well understood set of mechanisms, as many neurologists would agree. The understanding of the brain probably only began in the mid-30s, at around the time of frontal lobotomies and other primitive techniques. Today's approach with antipsychotics isn't all that much better, but it is a lot cleaner. Pills are used for everything from bipolar depression to autistic spectrum disorder (ASD). Drugs for the brain started the deinstitutionalization movement and moved sick people back into family homes, and many to the streets. Kids to adults are all using brain medicines, probably a lot more than they should. For example, teenagers aged 13-17 with bipolar disorder have their own specific medication, an action unheard of 20 years ago. It is widely understood that teenagers and puberty combined produce a volatile mix of emotions and moods, all of which tend to diminish over time. If the medical experts can understand teenage serotonin imbalances it is because they have spent a good amount of time studying the chemical exchanges inside the skull.

After scanning the neurological paraphernalia, I have identified that the neurotransmitter *dopamine* and the dark pigment *melanin* (and not the hormone melatonin) play an important role in our criminal game here. There are likely quite a number of interrelated factors and processes, but I am choosing to keep this medical investigation as tight as possible in order to prevent us from losing focus, and to remain well within my own amateur capacity for neuroscience. I am relying on the specific mannerisms I can observe right now, and in fact anyone can observe these two items if they watch the video evidence—the catatonic stupor of the large-eyed Baucus (4 blinks/min.) and the rapid-blinking of Lieberman (120 blinks/min.) and Pelosi (60 blinks/min.). All of this visible evidence is (at least) connected to dopamine and melanin. It turns out that both these constituents can be traced to a biological black box (nigra) inside the brain.

Dopamine is involved with cognition, motivation, sexual gratification, memory, learning, rewards and 'voluntary' movement. Melanin is a photosynthetic pigment that has an electroluminescent property. Melanin is considered to be an organic conductive polymer. Electroluminescence has led to the development of flat panel displays, LEDs, solar panels. Both dopamine and melanin are produced in the structure *substantia nigra*, which happens to also be our chosen black box. A high level of dopamine leads to schizophrenia and a low level of dopamine leads to Parkinson's disease. Each of these diseases is marked by the loss of control, involuntary movements, and odd behaviour. According to neuroscience, Parkinson's is not necessarily a genetic disease. It has environmental factors. Dopamine problems usually result in a diagnosis for a mental illness. And a decrease in melanin in the substantia nigra leads to lower dopamine levels. Certain herbicides (weedkillers) and pharmaceuticals, for example, contain compounds that reduce dopamine levels in the substantia nigra. Some neurotoxins can be converted into molecules that reduce levels of dopamine and herbicides and pesticides have been linked to causing Parkinson's disease. A biological weapon, for example, could also lead to a damaged nigra and to a whole host of neurological disorders.

Rapid eye movements are known as a *saccade*. A saccade is generated by a neuronal mechanism that by-passes time-consuming circuits, activating the eye muscles directly. The region in the brain that is involved with *saccadic eye movement* is again the *substantia nigra*. It is a brain structure involved in movement and a place in the brain where there are high levels of *dopaminergic neurons*, these are dopamine producing neurons. So there is a region in the human brain that not only produces dopamine but also can 'short circuit' access to eye movement. High dopaminergic transmission, too much dopamine activity in the brain, is linked to psychosis and schizophrenia. Antipsychotic medication (antipsychotics) inhibits dopamine at the receptor level.

The rapid eye movements of the two blinkers easily pass for a saccadic eye movement, for if these speed blinkers were doing so voluntarily it would have been some kind of childish prank, and that brings in the kind of essential information we need to approach this altogether speculative discussion. What is at stake here? I have decided to discuss a set of anomalous human characteristics that I found in the media, and I have presented my intuitive statements early on, but now I need to ground some of my thinking or risk losing the argument. It is a difficult argument with such little evidence and a restricted sense of observation, and having had more time to think and examine it I could further elaborate on the mystical qualities of my discussion. The rapid eye movements, because they are primarily involuntary, make it evident that they are not consciously controlling their high-rate of blinking. I have practiced blinking at their speed in the mirror. It is very tiring and requires a lot of focus. I could not, for example, blink at that rate and maintain any level of coherency during an important press conference let alone recall my speech on camera. In addition, it would become painfully obvious that what I was doing was on purpose.

Our blinking white-haired statesman is clearly accessing the substantia nigra because his blinking indicates a rapid flow of dopamine, a chemical release that appears to be uncontrollable and fits in easily with a saccadic eye movement. Both dopamine and saccades originate in the nigra, once again highlighting our black box. Could the nigra be acting as some kind of communications interface? Every brain has a substantia nigra. Does an android's nigra come better equipped with dopaminergic neurons and the photo-exotic melanin? There is no doubt that the nigra activation does lead to excessive blinking. This entire assertion is based on the following understanding: Career politicians would not act in this manner on purpose.

There is another possibility: Is Lieberman taking antipsychotic medication to reduce high levels of dopamine and then experiencing a saccadic eye movement as a side effect? Or, is he suffering from depression and boosting his dopamine levels and experiencing excessive blinking from a surplus of dopamine? We don't have his medical records, but we can infer that dopamine is involved here because we know that the substantia nigra is a major resource of dopamine and is involved with saccadic eye movement.

Both Lieberman and Pelosi have an eye movement condition particularly when they are situated in a similar position in relation to a political speaker. What is not clear is why it is selective. Why only when he or she is on the left rear flank and not when he or she is elsewhere or when he or she is speaking or experiencing emotion? Why the interruption? Does he or she have a form of schizophrenia and is being heavily medicated? Shouldn't they be on medical leave? Pelosi is involuntarily blinking during the entire States of the Union, and they average 60 minutes each. This is marathon blinking by any other name. She could enter the Olympics of Eye Movements. Ready, set, blink. Do either of them have another kind of neurological disorder, perhaps a more obscure condition like Huntington's Disease?

Huntington's Disease is a rare neurodegenerative disorder that compromises muscle activity and cognition. Among its many symptoms—jerking movements, mental disability, dementia, compulsive disorders—is included abnormal involuntary movement. An involuntary movement, in any of the neurological disorders I looked at, described the movements as irregular and not repetitive or rhythmic. In studying the video recordings closely, it is clearly noted that both Lieberman and Pelosi are blinking in a rhythmic motion like two fluttering wings on their faces. Additionally, in other video recordings and CNN interviews, for example, they do not have these ocular ticks, in fact, they appear to be quite regular, and by all rights normal.

The other thing that affects dopamine levels is drugs. Amphetamines and cocaine, for example, multiply dopamine levels causing temporary psychosis. Is there some drug issue in which causes their eyes to move wildly during press conferences? Perhaps there is a Press Conference Dopamine Disorder (PCDD) that no doctor has heard of till now. That still doesn't explain the rhythmic motions since even a drug interaction would produce irregular eye movements, and my observations, having examined the blinking phenomenon over and over again indicate that their eyes appear to have a consistency to their motions and are only interrupted through a brief expression of emotion or talking. In other words, the blinking is a default action during these high-profile scenarios and neither does this involuntary blinking interfere with their reactions to their surroundings. They are not inaccessible as would say a person with mental illness. Even Baucus, with his statue-like qualities, can maneuver with ease into full speech mode, without any loss of tempo or tone. It is as if the blinking is neither voluntary nor involuntary, rather it is something we haven't seen before, at least I haven't as a completely bottom string neurobiologist.

If it is a form of schizophrenia (in general as there are many forms), it is very serious because schizophrenia involves auditory hallucinations, bizarre delusions, disorganized thinking, depression, anxiety, and odd religious worship. We are not saying Lieberman (or Pelosi) has schizophrenia[3], but his rapid eye movements indicate high levels of dopamine which would put him in the category of schizophrenia, a disease characterized by too much dopamine, euphoria, alertness, over-confidence. Or it is also possible that they have Parkinson's disease, are short on dopamine, and are taking the synthetic dopamine *L-dopa* (laevodihydroxyphenylalanine) to boost their levels of dopamine. But then why aren't their blinks

---

[3] I cannot make any medical diagnosis since I am not a medical doctor. Certainly, schizophrenia is a very complex neurological disorder and many times people are misdiagnosed.

consistent in other televised situations? Most of their other videos, though not all, portray them as normal people. Why was it that during the deception-laden Wall Street bailout that Lieberman's dopamine is shooting wildly? And why is it that Baucus, who is later found flanking President Obama, is accompanying Lieberman? Did they require two robots to get the financial rescue plan passed? Counting Pelosi would make three. All three of them were actively involved in TARP (Troubled Asset Relief Program) and all three of them happen to exhibit artificial characteristics.

Again, we do not have a full study of these people. Ideally, we would want to monitor them for a week or two, and then to measure these observations against our speculative analysis. But that isn't realistic. And if they can control these organic mechanisms, by controlling the flow of dopamine, they could avoid detection even in our presence. We'd have to see them for a period of months, in all manner of situations, to really understand all of their natural ticks, physical nuances, and range of emotions. In the meantime, we rely on my amateur analysis.

Since dopamine is connected to involuntary eye movements (saccadic eye movements) then we know that dopamine is being processed (or is *flowing*) because dopamine is the fuel to generate eye movements. At this early stage, we can infer that the artificial person is being accessed by way of the neurotransmitter dopamine, as if perhaps an android could be molecularly controlled by way of a single neurotransmitter. A neurotransmitter is essentially a data signal with a set of instructions, a key. And male or female, it doesn't seem to matter. Each of them can disconnect, or stop, from the blinking dopamine connection by experiencing an emotional response. More interestingly, the saccadic eye movements connect both of them to the substantia nigra.

The substantia nigra sends a signal, cells release dopamine into striatum, the *basal ganglia*. The ganglia is a motion control center and is a place to direct body movement. Basal ganglia disorders include Parkinsonism, Tourette's Syndrome, ADHD, Obsessive-Compulsive Disorder, stuttering, and autism among others. These disorders are also found in the mysterious Encephalitis lethargica disease found later in the book.

So the nigra sends a signal to the brain to coordinate movement. The nigra is our interface and is the same no matter what race, color or creed. You have one interface to control many bodies. This is similar to a steering wheel on a car. No matter you drive a hatchback, a SUV or a sports car, each of them has the steering wheel interface. If you know how to steer a vehicle and drive its other components, you can drive the car. If the substantia nigra, our biological black box, is a chemically-based steering wheel then an advanced programmer need only to influence (or stimulate) the chemicals on this interface to input the data and then the onboard circuitry will take the necessary action. Incidentally, this might explain how a god archetype can produce a multitude of religions by influencing the production or inhibition of specific neurochemicals.

What is interesting about the presence of dopamine is that the blinking and the dopamine might only be a consequence of another action. If you drive a car and step on the gas pedal, the consequence is *acceleration*. But acceleration is also required in order to get onto the highway. It is possible that the dopamine is flowing for some other purpose and that the blinking is an indicator that there is information now flowing through this unit. On the surface, it appears that Lieberman and Pelosi are downloading information, data waveforms, from the cult classic *The Matrix*, without exaggeration, and then distributing that outward. Outward where? The only likely candidate is the speaker. And the speaker is speaking into a camera, a gateway into a network of television screens, and there are millions of homes with their TV sets turned on. Could there be some kind of multidimensional data transmission and the speaker is being used as a conduit? What I'm suggesting is that in addition to the wonderful rhetoric and the fancy suits and the blinking, could there in fact be another level to this charade, happening on a dimension outside of our perception?

Clearly, the speed blinking Lieberman and Pelosi are not in psychoses, they are not having a schizophrenic episode, but given the amount of dopamine flowing that should indeed be the case. They should have had a meltdown, but they didn't. Now if the dopamine, and its accompanying energy, were being channelled into the speaker because the speaker was acting as a conduit for energy distribution of some multidimensional quality[4], then the dopamine would not have a chance to overtax the individual and to send them into a psychotic episode. And the dopamine would be fueling some other kind of advanced invisible data transfer.

Additionally, as these blinking persons are artificial, it is likely that their internal machinery is built differently than a 'regular' human. That the neurological chemicals might be being used to run some internal machinery, like a mental turbine, or a molecular modem, and the blinking is an indicator that the turbine of influence is cranked on high. This could be a major difference between man and machine, their internal machinery utilizes chemicals and molecules far differently, probably far more efficiently. If we look at it from a technological view, say a piece of computer hardware, and thought of the brain as a kind of modem, we could say that a "human" brain modem was equivalent to a 56K dial-up modem via phone line and an "android" brain modem was equivalent to a high-speed cable modem. The difference is bandwidth and transmission speed.

---

[4] A multiple dimensional action can be any action that is outside of human perception or is visibly inaccessible. For example, light is a multidimensional quality. We can see a certain frequency of light with our naked eyes, but sunlight is composed on many other frequencies such as ultraviolet rays. Frequencies outside of the naked eye are occurring on another dimension, be it another frequency or vibration. It doesn't matter. Also, thoughts are taking place on other dimensions. We cannot see thoughts, we cannot measure thoughts, but we all have thoughts because we know we think. Thought is a multidimensional quality.

It could be the case that the speed blinkers are high-speed molecular modems streaming multidimensional data across visible space to a speaker in order to distribute it to all those with their television sets turned on. If so, if that outrageous hypothesis is correct, then that would mean that whoever is behind this incredible game is uploading some kind of information into everyday hardworking people. That information could be genetically-based and could be profoundly altering human decisions, human motives, human capabilities and human progress so that it is in line with the architects behind the wall. Of course, this is highly speculative.

The nigra interface on these two older models might be limited to a certain amount of data and the blinking could be a result from an older operating system, older circuitry and slower processing capacity. Certain computers blink more than others and certain laptop designs overheat more easily than others. The older android models are in positions of power and that is very important when you want to influence millions of people. They can't just insert a 21-year old Speaker of the House and get away with it. They needed to activate an age appropriate android.

The Large-Eyed Neo

Let's turn briefly to our other subject, the anti-dopamine unit. Perhaps Baucus can help us with our dopamine information-streaming hypothesis. This gentleman has an interesting stupor, a catatonic blank look on his face. A masked face. In Video#2, it is evident that he has a blank expression, a statuesque pose and a set facial expression. There is no rapid eye movement. He is barely moving, as if tripping out on some hallucinogenic agent. Does he have a form of catatonia? Doesn't the President of the United States mind that a catatonic person is flanking him and all of it is caught on live TV? The White House does monitor and measure the media appearances (and successes) of its Commander in Chief. Apparently, I was the only one who was watching.

Catatonia is a type of schizophrenia, associated with bipolar disorder and with some autoimmune disorders. Now I am not a doctor, so bear with me, sometimes catatonia is a reaction to medication. Catatonia involves an unresponsive, rigid stupor. It is a symptom of schizophrenia and a result from a dopamine surfeit. So we have the dopamine connection again, even though our second example is unblinking.

Catatonia (in our case unblinking) is produced from a *shortage* of dopamine, the blinking from an *excess* of dopamine. This neurotransmitter is playing a key role in the control of these artificial organisms. When a person has insufficient dopamine in their brain, especially in the substantia nigra, they develop Parkinsonism, the absence of dopamine. Might the senior politician have forgotten to take his L-dopa? Is there some other medical issue to explain his Reptilian stare?

Let's review some of the Parkinsonian symptoms in general: a fixed stony stare, deprived of movement and speech, strange eye positions, staring attacks, freezing, dazed, depression, restlessness, tremors. A reduction of dopamine can produce these symptoms and are usually identified as a neurological disorder. The large-eyed Baucus on TV, closely flanking the US President, has many of these symptoms, but we have seen him in Video #1, his speech is fine, he is coherent, and has a strong voice. Could he have a dual identity, one has Parkinson's and one doesn't?

In my three samples, we have two people with excess dopamine and one with insufficient dopamine. We theorized that the high-speed dopamine users are streaming some kind of information into the broadcast system likely through the speaker conduit. If that is the case, what is the catatonic Baucus doing without any dopamine? If dopamine is being used to alter reality, so to speak, what is no dopamine doing? It is possible that this catatonic unit is shut off, in sleep mode with the material reality, and is in sync with some electromagnetic grid system. He might be so in line with the artificial construct that he appears to *not be moving*.

What happens in a movie when the protagonist achieves the killing blow? The picture slows down to a crawl, the operatic music plays, the frame-by-frame scene is dramatized, close up of the eyes. What if that is what Baucus is in the midst of doing? He might be going frame-by-frame, hooked into the Wachowski Matrix. For what reason? Is he altering time-scales? Is he protecting the speaker, providing some (mind) shielding? Is he reallocating the time-space continuum so as to protect against some foreign attack, or to heighten the words of the speaker? These are some very profound notions, are they not? We are witnessing three very profound individuals. We might as well be working alongside Neo, artfully played by Keanu Reeves. These three people could be programs, holograms, disguised as robots *disguised* as people. Although the illusory world is another one of my interests, we have more than enough of the inconceivable to ponder here.

I feel it necessary to consider the possibilities in order to fully explore this once in a lifetime opportunity. I don't know if I can catch them again so clearly. It hasn't happened before. Ever. The bailout aftermath caused a very serious concern and might have forced the people behind the scenes not to follow the regular protocols. Had they followed the protocols, I might not have caught the nuances in the performance. I might not have noticed the slight variation in the vibration. Examining the evidence now, clearly there is something wrong with the picture, and this clearly is not a movie. Or is it? Is some advanced race making an advanced movie and we just happen to be in it, thinking we are real when we are merely actors playing out the roles they themselves wrote into the script? It could be.

We make movies in the real world. We hire actors. We set up cameras. We have scriptwriters. What if real life were a movie? Is that what I witnessed? Perhaps I was witnessing a scene in the life movie and it was then I glimpsed at the set and noticed that there was a cameraman and a director behind the invisible wall. It is highly likely that the androids, given their programmable state, are perfect actors. That you can program them according to what you need. You need to sell a war, upload the war script. You need to sell healthcare, upload the healthcare script. You need them to procreate, upload the sex script. A robot acts according to its programmers. If these three people are robots, then there is a programmer somewhere. There are technicians. And they know how to use dopamine flows to remote control their robots. Furthermore, the same architects also designed the nigra interface and purposely engineered an excess of photo-exotic melanin.

We can better understand the contrast between the speed blinking Lieberman and Pelosi team and the nonblinking Baucus, between the flow of dopamine and the interruption of that neurochemical. The spectacular involvement of the substantia nigra, still a minor section in the brain, is, oddly enough, a substantial piece of evidence to our crime scene investigation. It might be true that especially the older android models, manufactured at the time of World War II, had prominent nigra features that could have been translated as having broadband existential qualities. Control by chemical agents, probably not new in the mind control business, enable the control of specifically placed automatons, a preferred method in clandestine mechanisms of control. It is a method of puppetry that vastly improves upon the traditional method of strings and pulleys.

DROPS OF DOPAMINE

Parkinson's disease is a result from an absence of dopamine. There is an estimated 500,000 dopamine cells in the substantia nigra. If you lose over 60% of these cells you will likely develop Parkinson's, a disease with no cure and no cause. Two famous people have Parkinson's, one is Michael J. Fox (actor) and the other is Muhammad Ali (professional boxer). We know them well.

Michael J. Fox   Muhammad Ali

If dopamine levels are connected to Senator Baucus and if this large-eyed political figure is an artificially created being, an android, if I can demonstrate that, it is possible, plausible that Michael J. Fox and Muhammad Ali are a lot more artificial than they appear. The difference is that the androids in my recordings are being maintained by a very advanced set of secretive people, because they are being regularly serviced, and our celebrities are stuck with 19th-century medicine. But that would be impressive—if what we thought was a disease was actually a misinterpretation of the type of person, a mistaken identity, seeing a biological person instead of an artificial person. Because if we were to realize that our celebrities were artificially produced, as impressive as examples they are, then we could treat them in a whole new way. There could be a whole list of artificial diseases, that is, diseases particular to people with synthetic DNA. Are there synthetically created people living among us? Certainly, this is a major issue, a subtext of this entire treatise.

The evidence of one or three synthetic people gives rise to the possibility of 300 synthetic people. We cannot say that the entire world population is synthetic, although that is a possibility, but we can infer that there is an artificial demographic that we never knew existed, verily a new race of people yet to be discovered. They might be sitting across from me at my coffee shop. A pale-skinned woman drinking coffee, reading a book (maybe even this book), thinking she is alive and biological when in fact her parents are machines of some advanced multidimensional quality. We could even catch the secretive advanced people and ask them for the cure for this Parkinson's because surely they would have it. Why is it that these secretive people refuse to show themselves? Isn't it about time in human history where we pull back the veil of secrecy and expose the technological underpinnings of life itself?

They created androids like in the Michael Crichton-directed film, *Westworld* (1973) and its non-Crichton sequel *Futureworld* (1976), the technicians behind the walls, why they get paid to repair the androids, to reprogram them. In *Futureworld*, the clandestine android makers were copying human leaders and journalists and swapping the artificial clones for the real thing, which were then destroyed. In this way, the world could be controlled by human-looking androids all loyal to the android makers. Even more impressive is the fact that this film is now nearly 40 years old. That's four decades of android integration and copulation that we unfortunately did not monitor.

Parkinson's disease is a degenerative disorder of the central nervous system. One of the reasons it comes about is because there is 'not enough dopamine' being produced in the mid-brain region. We are familiar with this mid-brain region—again in the substantia nigra. People who recovered from *Encephalitis lethargica* (sleeping sickness) in the late sixties developed post-encephalitic Parkinson's disease.

Parkinson's – not enough dopamine.

Schizophrenia – too much dopamine.

Both Parkinson's and schizophrenia involve the neurotransmitter dopamine. Catatonia is a symptom of both and can also come from a drug side effect. Catatonia is also involved in autism, a developmental disorder that has been rising in recent years.

The Mysteries of Encephalitis Lethargica

When we study Baucus, and his blank stares, his statue-like condition reminds us of *Encephalitis lethargica*. It is an atypical form of *encephalitis*—brain inflammation. *Encephalitis lethargica* attacks the brain and leaves people in a statue-like condition, speechless, and motionless. There was an epidemic between 1916 and 1929[5], it claimed nearly a million lives, froze millions more, and ended just around the time of the Great Depression. We are going to remember the Great Depression because it is an important part of the overall

---

[5] I have found a number of date discrepancies for the Encephalitis lethargica outbreak. The general agreement is a 15-year period between 1915 and 1930. It was believed to start, the first patients, in Vienna in 1916 and it was believed to have mysteriously disappeared by 1927 or 1928. Again, this disease has not only an air of mystery but a true lack of medical inspection. It is a forgotten disease yet I see no reason to forget such an important event in human history. Five million people were infected with a disease that had a 50% mortality rate and directly followed the 1918 Spanish Flu. Contrast this to the Holocaust which has many books and films that account for it, so that no one forgets. Yet Encephalitis lethargica is conveniently forgotten. It begs the question: why? Why is it so easy to forget? In my fringe medical investigation, I have found direct links between encephalitis lethargica, Parkinson's and the androids. I did not expect nor intend to find these incredible connections. But on a neurological level there are fundamental connections. Not only that, but the period in history that the pandemic occurred strikes a large bell with the historical pattern of these advanced races of people I have mentioned, thought I do not have their photos.

discussion. The survivors of encephalitis lethargica sank into a semi-conscious state that lasted for decades. When L-dopa was invented, it was used in the late 1960s to revive these people who had been sleeping for decades. L-dopa is also used to treat Parkinson's with mixed results. The key difference between encephalitis lethargica and Parkinson's is that nerve cell damage for encephalitis lethargica in the substantia nigra, the control center for voluntary movement, is completely depleted, leaving the afflicted in a motionless state.

A 1990 American film, *Awakenings*, based on a 1973 memoir of neurologist Oliver Sacks use of L-dopa to revive the survivors of encephalitis lethargica starred Robin Williams and Robert De Niro. The patients in catatonic states 'woke up' for a short time becoming normal. But the results would not last.

Notice the dopamine connection again with the statue-like condition, a condition that one of our key examples seems to demonstrate. It is clear on the recorded video that this person has some kind of condition, but, this condition is selective, in other words, he can control his disease. He can control his statue-like state of awareness. He is able to *switch* from one state to another state. This appears to be the same for all three people. By normal medical definitions, we have no control over serious brain inflammation or serious dopamine imbalances. And yet all of our examples have *control* over their disease symptoms.

The white-haired Lieberman can stop blinking when necessary; Baucus' catatonic posture can wash away when necessary, the blinking Pelosi stops blinking when she has an emotional response. Could it be that Pelosi, when in 'android mode', has no empathy and can access her mechanical abilities and as soon as she allows empathy to flow it shuts off (via dopamine) her android technology? It is like they can switch from android to humanoid and from humanoid to android. Switching when necessary, without any difficulty.

I am going to ask the unobvious question: if the individuals I have identified in my recordings, three political figures based in the US, three example of many that I could have chosen, if these individuals are androids, artificially-created people, then it would mean that there are indeed more androids in the world. The androids use dopamine as a central existential currency. In other words, to access their android components, the android needs adequate dopamine levels. A shortage of dopamine can cripple the android, can turn them off, literally.

What modern medicine labels as brain inflammation, encephalitis lethargica, could in fact be some interruption in the android programming. Sort of like how the androids in *Westworld* started to malfunction from a computer virus. It would be logical that the androids can malfunction and that they need maintenance. Of course, only VIP androids, such as our political samples, are getting the attention because Parkinson's and other incurable diseases have infected many people.

So here's my question: what if the technological masters could create programs, like computer viruses, and deploy them into society creating an epidemic that strategically targeted a certain class of androids, or androids with a certain frequency of operation. They could attack androids because no one in their right mind believes in androids. And the conditions would be interpreted as brain inflammation, and labelled as encephalitis lethargica for example, when in fact, these are android-specific synthetic weapons targeting other androids who *think* they are humans. And our missing link is the neurotransmitter, dopamine. It is rather remarkable to think that at least some incurable diseases, including Parkinson's and encephalitis lethargica, are not human diseases and instead are android-related illnesses. This is probably why the medical community has yet to find a cure, because they are treating a synthetic illness with a biological pathology.

Dopamine is our link to the androids in my selections and if I have indeed selected androids 'then' dopamine is our link to other members of the population, including people like Michael J. Fox and Muhammad Ali. I can now infer that Mr. Fox and Mr. Ali have portions of synthetic DNA and that their genome has been engineered at one time or another. The presence of Parkinson's is proof that they are not entirely human. By the same token, they are not necessarily 100% synthetic, as we are defining here, since that might have completely crippled or killed them outright. And I am inferring this because that might help us find a faster cure for their artificial disease, their android specific disease. If you treat a tiger's disease with the cure designed for a rabbit, you may not be as successful as if you treat a tiger's disease with the cure designed for a tiger. Ancient Chinese saying.

Is it possible that the processing of dopamine is connected to the access of android functions? More specifically, is dopamine the key to starting up and driving the android? Is that why bipolar disorder is so prevalent? It reminds me of the fuel injector in an automobile engine—if the fuel-to-air mixture isn't right, the engine doesn't run smoothly. What if when the dopamine level isn't right that the android lacks their entire functionality and only has access to its basic functions, functions that are human-grade, and if the other chemicals are out of balance, a flush of dopamine could lead to aberrant and dangerous behavior. In other words, the activation of the brain via the dopaminergic key does not guarantee in any way the productivity of the robot.

The flow of dopamine might also close the channel to some mainframe computer, denying the stain of the immaterial world. The absence of dopamine then directly opens the channel to the immaterial, to the multidimensional architecture which cannot be seen. A reduction of dopamine might make a human more spiritual and an excess of dopamine might make the human more materialistic, more productive. No dopamine could produce the benign hallucinations that

Dr. Sacks spoke of in his memoir on the post-encephalitic patients under his care at Mount Caramel.

> "They hallucinate to survive—as do subjects exposed to extreme sensory, motor or social isolation; and for this reason, whenever I learn from such a patient that he construct a rich and benign hallucinatory 'life,' I encourage him to the full, as I encourage all creative endeavours which reach out to life."
>
> *~ Dr. Oliver Sacks*

We know that many people on the planet are "waking up." What if waking up involved dopamine processing as a major component? What if we access our machinery by managing dopamine? If that is the case, then anything that affects our dopamine levels is preventing us from "waking up." The question then becomes: what kinds of things affect our dopamine levels? The answer is quite ugly. Dopamine is involved with pleasure. Dopamine is involved with consumption and addiction. Dopamine is involved with rewards. There are three very powerful dopamine antagonists, enemies against dopamine stability—food, sex and drugs. Wham! Food, sex and drugs. Obesity, prostitution and pharmaceuticals. There are more diet books in the world than ever before but obesity is at epidemic levels. Sex sells and it is what advertising is largely about. Sex in movies, in magazines, on the streets, in the churches. Drug wars and pharmaceuticals drugs are big business. Our dopamine is in big trouble.

How are we to wake up when our dopamine is out of balance? Even you can produce more, there are more temptations. And if you produce too much you go into psychosis, you get the symptoms of schizophrenia. And if all the temptations consume you, you end up depleted and with Parkinson's or similar. Most people with Parkinson's are over 60, all of our examples are over 60. And then there's Autism, a new spectrum of developmental disorder which we don't cover but have some odd behavioural characteristics. Might it also mean that autism is an artificial genetic disposition?

If the human body requires an adequate amount of the neurotransmitter dopamine or the body cannot move properly, then we could also say that dopamine is data, the kind of data to instruct movement. Our blinking examples prove that there is dopamine present in their system. Dopamine is produced in the nigra structure. Some external energy is being applied to switch these people from one identity which is independent to another identity which is subservient. The nigra controls voluntary movement and their blinking is involuntary. And the black pigmented nigra is loaded with melanin and we know that melanin is an organic switch because it has high electrical conductive properties.

Android operations 101

We have a machine and we are learning to understand how that machine operates. We used a few criminal clues to dig inside the human body. And we did that to understand how an artificial body operates. I identified a structure in the brain, substantia nigra, as a location in which to remote control the mechanical vehicle; otherwise known as the body. A better term for substantia nigra is the 'Biological Black Box'.

Then we have the body itself and we know that the human body is a chemical body. Chemical processes and molecular exchanges determine the essential life forces of an individual. Chemicals produced in your brain determine whether you are a good student or a bad student. Chemical eggs in the female determine her child bearing ability. Protein synthesis determines the strength or disease of the individual. The human is an organism but that organism is chemically ordained. This is an important side note because I have identified two instruments. One is dopamine, a neurotransmitter. The other is melanin, a black pigment. If we have said that melanin is an organic circuit that works as an effective 'switch' then dopamine is the chemical key that activates that switch. We have not a metal key or a computer key; we have a chemical key; a molecular key: a quantum computer key. Our key is a neurotransmitter produced in our biological black box. And we have an electroluminescent switch. Our quantum-based brain structure. A quantum ignition control system. Dopamine and melanin. And all we need to apply is a little energy. A small voltage. A charged spiritual pulse to control the biomechanical vehicle.

If the soul is nonbiological and we know that it is then made of energy, we now have our energy source to use the dopamine key to turn our melanin switch in our biological black box. We now have the basics of a quantum robot disguised as a human being. And we similarly know that if a substitute multidimensional energy (eg symbiote) accessed that dopamine key, it could hijack the vehicle, especially a vehicle whereby the driver didn't know what was going on.

Look at the automobile: You need a driver with a key and the key is inserted in the ignition which involves a switch. By turning the key in the ignition, the energy stored in the battery activates the starter motor, sends electricity to the spark plugs, the crank shaft rotates the pistons and the engine turns on. Once the engine of the vehicle is on, you still need a driver's licence or the ability to drive. If you don't know how to drive and I give you the keys to a Ferrari, you will not be able to drive it.

Do the androids have souls? Do they have synthetic souls? Every machine requires an operating system. We haven't been able to investigate the consciousness of the people but we can confirm that they are conscious, they have intelligence, they have a voice, and as we've said they have families.

To remote control these android machines, you need to not only know how to operate the biological black box but you have to apply energy in order to access it. You could also apply a wireless energy like a telecom network. The phone has its own battery but it literally comes to life on the network. We haven't talked a lot about the external source of energy. We know that much of this technology exists on another dimension and we can infer that it is a dimension that isn't far from the soul because our wireless energy has to be made of a similar quality of energy of the soul (or consciousness) because otherwise we might damage the body or the body won't work.

We briefly discussed an artificial network that could be operating a level of energy that is in line with the soul. But we have to remember that the soul is a nonbiological component to these advanced people. The soul is a spectrum of electromagnetic energy that exists on a dimension that, till now, has escaped us. Probably because we are stuck in the 3D world of limitation. And the advanced people, the programmers creating androids, are either multidimensional, or just really really smart. This will be something for a future discussion.

CREATING A SLEEPING SICKNESS

There is a very ancient weapon that has existed for thousands of years. It is called *infectious disease*. From the plagues of Egypt to the pestilence of ancient Greece, infectious disease has remained a powerful weapon against mankind's progress. Particularly interesting is *Encephalitis lethargica*. It is particularly interesting because it produces Parkinsonian symptoms and Senator Baucus has Parkinsonian symptoms—the statue-like presence, the reduced expressions (frozen) in his eyes and face, the slowed response. Of course, I am not saying he has the disease as much as I am saying that being artificial can put the person in a state that looks disease-like. This is a vital distinction.

Baucus could very easily be perceived as having Parkinson's disease in Video #2, but we have seen him speak fluently in Video #1. He had no problem speaking. Was he under a different medication at that time or does

Parkinsonism have a connection to Androidism? It's a fundamental question and one that may not be answered properly for quite some time. Furthermore, the transformative qualities that Baucus demonstrates in the video sample, and in the many normal sample clips online, is indicative of a man in control and therefore unlikely under the influence of some drug. And should it have been a Parkinson's-related disease, a dopamine deficiency in the nigra structure, then, most likely, a 'la Fox and Ali, he would not so easily be able to normalize his symptoms. In studying Muhammad Ali's interviews after Parkinsonism took effect, I can see that this former professional athlete cannot 'normalize' his condition. He is a prisoner to its effects. Baucus cannot be under medication or suffering from Parkinson's, even though we might perceive him as such, and if not then why the anomalous behavior?

What if as an android's dopamine shifted from abundance to shortage—the android could control their type of behaviour or interaction with the environment. An excess of dopamine, and perhaps other neurochemical combinations such as acetylcholine and serotonin, would allow certain androids to aggressively interact with the reality construct. Shutting off the dopamine then would connect the android to a deeper aspect of the reality. What is depicted in the video recordings is that all three androids can shift in and out of states of awareness (altered states). For example, they can be blinking rapidly, as if transmitting data to the other human terminals, and then by way of a simple emotional reaction they can stop blinking and return back to a normalized disposition. In other words, they are in control of their disease which leads us to believe that it is not a disease. And yet it is not normal. What is it?

It is what I consider to be a different type of human being, one that can switch between a human mode and an android mode, and in human mode, by all definitions, they *appear* human, but in android mode they *appear* out of character, perhaps even symptomatic. What impresses me about all of this is that if you measure them as a whole person you cannot see the human-android dichotomy. Only by isolating the android moments, if they are significant enough, and not without observational error, only then can you justify whether or not they are of a synthetic nature. Then it is also a matter of deciding to what extent they are synthetic since it is also the case that not all androids are 100% android.

In order to live among and to pass as human beings it is in the genetic designer's interest to incorporate some indigenous human DNA into the genomic data. It only makes sense. This is also why the Lieberman sample is perhaps a better example of a pure android because I think his synthetic DNA ratios are highest among all three of them. I would rate Baucus as second and Pelosi as some other kind of model entirely. Pelosi is of particular interest in this android investigation—I feel like I'm in a Ridley Scott movie right now—she has a significant amount of synthetic DNA, as described by her one-blink-per-second blink rate (an average person has less than one blink every two seconds), but she also has an amazing ability to lead. Under her authoritative control, Pelosi was able to gather all the political minions on Capitol Hill and to coordinate a financial coup, essentially the economic crime of the millennium. The fact that she is a woman has no bearing on this outcome, if anything it works against her in a misogynistic Congress. The fact that she is synthetic gives her certain advantages. There is some X-factor that gives her this commandership.

In the theatrical business, a thespian trains for years to be able to slip in and out of character, switching from an angry emotional outburst to a resolute calmness with relative ease, whatever is required from any particular scene. What allows them to do that is acting skill: a talented actor can do it all that much more naturally. A great actor can do it and make it seem normal; an amateur makes us laugh because they cannot hide the illusion. It is an illusion, the illusion of impersonation and movement. This is what acting is all about. It is about lying to the audience in such a way so as to tell a story, and the audience will only watch if they can believe that this could be real.

These three artificial persons are seamlessly going from Parkinsonism, or similar condition, to normal and from REM in waking state, with rhythmic eye movements, to normal. Yet, as much as they are politicians and somewhat actors, they are not professionally trained thespians. Plus, if they were created then their actions would not be independent of their programmers. That is to say that during these important conferences, a time when the audience needed a nudge in a particular direction, it would be logical to conclude that the programmers were involved in their puppet's lives. It is logical to think that the puppeteers were remote controlling their puppets. We confirm the presence of excess dopamine in the blinking examples. Similarly, we confirm the lack of dopamine in the masked example. All of this can be accomplished by the puppeteer by using dopamine as a key to drive the artificial pod. It wouldn't be unreasonable to think that. They were steering the vehicle with the substantia nigra interface (ie android steering wheel) and using molecular magic to entertain the unsuspecting, and uneducated, general public.

The sleeping sickness epidemic from 1916 to 1929 infected 5 million people, and killed at least a million. Those that survived were left crippled and probably comatose for years, decades even, and the rest had Parkinsonism. And the brains of those afflicted with encephalitis lethargica had damaged areas of the midbrain which we identified as the *substantia nigra*—the very exact same midbrain region that not only produces dopamine and is rich in melanin, but is also behind schizophrenia and Parkinson's disease. The nigra structure also is identified in encephalitis lethargica, a mysterious disease that appeared mysteriously and disappeared even more mysteriously at the same time of the 1929 Great Depression.

Going back to that period in history, in 1918, the Great War was settling down (1918 Peace Treaties) when suddenly the World is hit with the Spanish Flu. The greatest enemy of all is a virus that within a couple of years will kill 50 million (estimates up to 100 million) people. World War I, with all the bombs and bullets, claims the lives of not even 10 million. A Swine Flu kills five times as many people as a war and it occurs as the Great War is slowing down.

The war officially ends in 1920. The Spanish Flu *morphs* into encephalitis lethargica, a global infection that cripples people. It produces Parkinson's-like symptoms, a mask-like face, loss of equilibrium, rigidity of muscles, loss of automatic movements, respiratory problems, catatonia, chronic fatigue, drowsiness. Those who don't die are frozen for decades.

The 1918 Spanish Flu was a H1N1 influenza A virus of swine origin. This flu strain would show up again in 1976 in America. In three months, 24% of America would be vaccinated despite only an isolated outbreak. The vaccination program would be stopped because the vaccines were found to be more harmful than the flu. It would later be called the Swine Flu Fiasco. In 1957 and 1968, Asian countries are hit with a H3N2 strain of the influenza virus, two to three million will die. Jump to 2009, the H1N1 shows up again. The World Health Organization calls it, without consultation or proper research, a pandemic, orders vaccinations for everyone or paraphrasing "the world will end."

Americans are reminded of the vaccine deaths in 1976, they don't believe the media-exaggerated crisis, something smells foul. The 2009 Swine Flu is believed, in conspiratorial circles, to be a biological weapon being used to force vaccinations that contain substances harmful to society. For example, the flu attacks healthy people and contradicts the normal pathology of flu outbreaks that attack weak immune systems and the elderly. Many people believe that the flu crisis is being generated to force people to inject a tainted vaccine into their otherwise healthy bodies. Truth speakers get active. The crisis is being manipulated. Sixteen thousand people would die, nearly 2,000 healthy children, as a result.

Interrupt the programming

In the *Discourse of Persons Artificial* and *American Androids* videos, I introduced the idea that the Great Depression of 1929 and World War II were invented by an advanced race of people, the same race of people who invented the Androids on Capitol Hill, three elderly politicians. These three people were created during World War II while many nations were trying to defeat one another. The pattern of the 2008 Wall Street Bailout, a financial stimulus that would put America on the brink of a Depression (or a prolonged recession in light of a more efficient monetary system) in 2009 with record unemployment and budget deficit, coincided with the 1929 Depression. In the United States, a number of Republicans including Newt Gingrich, seizing on the opportunity to strike against the Democratic Party at the helm, have stated that the economic situation facing America in 2012 was the worst in 73 years, which happened to coincide with the launch of World War II.

In both 1929 and 2009, the Federal Reserve Bank, with the power to control money supply, played a central role. Ironically, the year 1913 was the time of the Federal Reserve Bank Act, created in order to prevent a financial collapse. Why in one period a depression occurred and in another a recession, I think, is because of the flow of money as a result of monetary transactions. Economics relies on a certain speed of monetary exchanges. The paper-based 1929 system in no way could match the microchip-based system in 2009, and as long as the monetary system did not seize up then it couldn't crash. What could've been a 2009 Second Great Depression instead became a deep recession (Economist Paul Krugman and Investment Broker Peter Schiff called it a Depression), which was likely a result of technological monetary systems such as online banking and shopping, two things that prevented an economic seizure.

The presence of three androids in the modern day meant that there was an advanced player in the game in the modern day. If these advanced cultures are 1,000 years more advanced than our scientists today then, comparatively speaking, they were 1,000,000 years ahead of the science of the 1920s—no microchip, no stem cells, no nanotechnology, no space flight, no drones, no credit cards, and no nuclear power. Why the people in 1929 were primitives in comparison to the people living in

2012. If the advanced races were 1,000,000 years ahead of them, we today are 1,000 years ahead of our great-grandparents. If there were androids committing an economic crime in 2008, and if three of those androids were created in the early 1940s, then you can bet that these geneticists were also around in the 1920s, which means that androids were in use as well. It has to be the case, since the method of command and control requires the android interface, that there were androids involved in the economic crash of 1929, only that we don't know who those people are, yet. In the meantime, we can employ logic to understand the extent of our enslavement. It is our enslavement if a band of advanced technocrats, well beyond our scientific ability, has been misleading us for a great many decades, and have once again tried to cripple America in order to induce some other kind of global catastrophe.

Further, by relying on my medical investigation and a dopamine hypothesis, I have been able to link the neurological conditions of encephalitis lethargica with the neurological conditions of my three artificial persons. And I suggested that the encephalitis outbreak from 1916 to 1929, ending at about the time of the Depression was an orchestration by these same mysterious advanced races of people because only they understood how to cripple the android population. Because only they were aware of androids among us; because they put most of them in place.

Now, we have a 1918 Swine Flu preceding the large outbreak of encephalitis and we also have a Swine Flu in 2009, at the time of a failing US economy. The repetitions are adding up already: H1N1 virus, dopamine damage, a Depression—one in 1929 the other (though never officially stated) in 2009. What is connecting these pieces are androids, fingerprints linking two events separated by 80-100 years.

Why does H1N1 kill 50 million people in 1918 and only 16,000 in 2009? One, people are aware of biological agents, they recognize the signature of a neurotoxin being applied to force vaccinate society. In 1918, very few would suspect biological warfare being purposely used by an enemy and being covered up by the military. The primary chemical used during World War I was mustard gas. We can guess that there were other toxins used more strategically. If so, a strategic attack could cause an outbreak of a strange and mysterious disease with a high mortality rate. In fact, it would be odd to think that a naturally produced environmental disease would be so lethal. It would make sense that a weaponized agent was used. The first wave of encephalitis lethargica was discovered in Europe after an intense battle between the French and the Germans. French soldiers got sick. Were the Germans using weaponized biological agents? What did they use besides mustard gas?

Also, at the time, the science of the brain, neurology, was very primitive. They were soon giving the mentally ill, frontal lobotomies. No one understood the dopamine connection with encephalitis lethargica, but autopsies showed that the dark pigment in substantia nigra had indeed been damaged. Is there a connection between the damaged substantia nigra of encephalitis lethargica and that of Parkinson's disease? If indeed there is a connection; and if during WWI the damaged pigment was a result of a biological toxin (a neurotoxin) then that tells us that Parkinson's is not a naturally occurring degenerative disorder, and that it has the kind of cause we are afraid to know about.

Not until 1969, when Dr. Sacks would apply L-dopa to these semi-conscious post-encephalitic survivors would we begin to understand how the brain works. Then you also have better genes, improved diets, stronger immune systems, what killed people 100 years ago doesn't have the same efficacy. It is efficacy that they wanted. It is no accident that these three things—Swine Flu, Dopamine[6], Depression—are present in 1929 and in 2009.

Our discovery of artificial people on Capitol Hill meant that some advanced race had already hijacked America, but that wasn't enough. Did they try to artificially

---

[6] We could also consider the other prevalent neurological disorders such as bipolar disorder, schizophrenia, ADHD and ASP (autism spectrum disorder). I haven't done so because each of these disorders is different. Their commonality is neurology. And dopamine plays a significant part, but there is a direct biological link with Parkinson's, Encephalitis lethargica and the androids on Capitol Hill – the *substantia nigra*. The artificial persons are all accessing their nigra structures, the encephalitis lethargica patients all had damaged nigra structures and Parkinson's patients have degenerative nigra structures. The substantia nigra, the nigra switch, is common to all three and all three occur from 1916 to 2011. It is a constant. I think that the substantia nigra is the interface that will link this physical reality to the advanced races of people who are in some other dimension.

interrupt human programming by introducing an infectious agent that would be tainted with a newly-purposed toxin so as to kill, cripple or put into coma a large part of the world? The 2009 Swine Flu was a pandemic, it launched a global vaccination program, it was a global attempt to kill people and if not kill them then to cripple them and if not to cripple them then to put them to sleep for decades. But were they targeting a certain class of people? Were they targeting other artificial people, a natural threat to order, who would be more easily damaged by their 1918 toxin? Are we seeing the presence of competing android models on the planetary market? Is there another group of android makers that has built androids for more benevolent and compassionate use?

You see the problem we are now encountering. The picture is becoming quite complicated. The pattern of behaviour, the fingerprints, the risk of economic collapse, the strain of flu, all of these were present in (circa.) 1929 and 2009. That, to my mediocre eye, is not a coincidence. It is an indication that the same architects are behind-the-scenes in both distinct time-frames, because their fingerprints are there, things that are so specific that they cannot have appeared *unpurposely*. And if a weaponized agent was used to cull a certain class of the human population, likely a synthetically-based subspecies of hybrids, then a newly-revised toxin was intended to be used to cull a modern group, only that there was a greater resistance than they expected and they managed only to vaccinate a small number of people.

We still don't know the repercussions of those injections or if some other agency acted to reduce the toxic effects. We do know that the architects are the same. Instead of killing off people with their toxins, they might have only been able to damage their bodily systems and have produced a wide variety of neurological disorders as a result.

The Parkinson's-like symptoms, as described on Senator Baucus, were also produced by the mysterious illness encephalitis lethargica. What killed or crippled people in the 1920s, some unknown biological agent, could in fact be controlling the Baucus unit. How can I say that? Baucus' demeanor is impressive not only because it is extreme, a dead stare, but more so because he is in control of it. The encephalitis sufferers were not in control. When they became frozen, they remained frozen. Contrast this situation with Baucus who can be completely frozen in movement, staring blankly outward, and he can shift out of that phase into a normative state of operation. I attribute this to an immaculately operating substantia nigra, one that can shut off and turn on dopamine without ill effect. But the encephalitis lethargica outbreak is a big concern here, not just because it came and went like a thief in the night, but why—because it specifically targeted the substantia nigra, the structure in the brain that gives an android an ability unlike any other. This also tells us another important thing and that is that the people who were afflicted, or died, with encephalitis lethargica must necessarily be made of synthetic DNA, and that those who died outright were probably more synthetic than those who managed to fall into an endless coma.

A KIND OF SYMBIOTE REVOLT

Encephalitis lethargica has some key pathological characteristics that urges us to redefine the nature of this mysterious disease, a disease that can appear and disappear with relative ease, an attack on the nervous system that disables the host (or kills the host), a disease that changes the disposition of the patient overnight, an invisible disease that has no identifiable cause or known cure. We have enough reasonable doubt, as we did with the androids on Capitol Hill, to step out of the box. We are going to step outside the box as before in order to see if reclassifying encephalitis lethargica can help us relate to our much more advanced other world; a world that includes an advanced race of people who can cloak themselves beyond our perception.

To start off, I am going to say that encephalitis lethargica is a good example of a multidimensional disease, it escapes cause and cure; therefore it is logical to think that it also exists inside another dimension and could be attacking the patient on other levels of awareness (and therefore explains the autoimmune attacks of the patient). This would not be unlike psychological and spiritual problems that have no physical identity or literal device and rather require delving into mental, spiritual and emotional dimensions, with unknown results. This time, I want to delve into technological dimensions since that is the basis of my treatise, we are fundamentally exploring new sciences and technologies in order to come to terms with the presence of persons artificial on TV.

The patient of encephalitis lethargica is subject to some kind of bodily invasion (as opposed to just the immune system attacking itself). The patient is a host to the kind of organism that can subdue its normal operations. This technological host doesn't stop there, it continues to feed off of the host until the host becomes increasingly incapacitated, or dead. The only reasonable, if fictional, concept I could find to help us understand sleeping sickness is a *symbiote*. A symbiote in comic book literature is *a living organism that bonds with other organisms in order to survive*. Quite often, as in the case in the Marvel Comics Universe, the symbiote is an interstellar life form, or extraterrestrial organism. This was seen in the Sam Raimi-directed feature *Spider-Man 3* (2007) with the villain Venom as well as the comic book series of the same character.

Typically, symbiotes take over a species they come into contact with and feed off of their emotions, their energy and even brain chemicals. The symbiote fuels the animosity of the host and then feeds off of the anger, the by-product of his production. The host often experiences rage and varying degrees of psychoses, not unlike schizophrenia. Only that the comic books are not scientific manuals and are purely entertainment driven.

Symbiotes

Symbiotes form tendrils that attach themselves to the host and take the host over, often giving them superpowers, but the symbiote can drive the thoughts of its host or even of a potential host. It is intelligent. In some comic versions in *Iron Man*, the symbiote is a virus-like biological organism, a Bio-weapon. But because the symbiote is created by the same technocratic administrators, it is a very advanced device, and it is invisible. The proof of its existence, so far, is Parkinson's, and a whole host of other autoimmune diseases which we haven't the space to discuss and something I have detailed in my book *Reality Medicine*.

Is it possible that the flu pandemic weakened the immune systems of enough people so as to allow a mass of technological symbiotes to invade a certain class of human? Is it possible that a multidimensional symbiote attached itself to the inner components of the host, perhaps at a chemical level, perhaps adrenaline, and then feeding off of the host took on a more physical form and attached itself more thoroughly. Pretty fantastic idea, isn't it? Of course, I am quite serious about this.

The immune system weakened by a planned infectious disease, the Spanish Flu, has given the weaker life form, symbiote, access to the advanced machinery of this planetary race because the symbiote is from elsewhere and the human immune system is designed to be resilient. The multidimensional organism deposits its tentacle roots into the host's nervous system, essentially hijacking the host, able to shut down the host. In order to protect itself, the host's program will attempt to shut down before the infestation is too far. If the symbiotic organism manages to break into the operational command of the host, the host will be shut down permanently, chiefly the respiratory systems. The result is death. The doctors observe the dying patient but they cannot see into the dimension of this entity. They see no cause and all effect.

As I have stated from page one, this entire discussion is a science fiction discourse because it relies on the presence of three androids, human-looking robots that I have directly observed, as many others have by now, on prime time TV and have recorded on my video camera and now have identified by virtue of neurological characteristics. Therefore, when speaking of a symbiote we have to realize that the symbiote is an artificial entity. Why must the symbiote that brings on the sleeping sickness be artificial? Because the kinds of people the symbiote prefers to attack are artificial. Because the world we live in is artificial. Take your pick.

There is more evidence to support my hypothesis, the sleeping sickness did not affect everyone. In fact, family members and others could visit a person who had encephalitis lethargica and not get encephalitis lethargica, for example doctors did not report getting the disease, and nurses did not report getting the disease. Why? The disease isn't contagious. The disease is allocated to the host. The disease, to me the nonphysician, is technological. The disease, to the physician, is mysterious. It entered mysteriously in (around) 1916 and it disappeared mysteriously in (around) 1929. It did so, as I will argue, because it is an artificial disease. And a more proper way to describe an artificial disease is a *program*. A viral multidimensional program designed to attack a robot host.

The big problem with our view of the world is our biological perception. We haven't been able to escape our biological chains. This entire discussion is fundamentally based on a *nonbiological perception*. It is founded on the presence of artificial persons on Capitol Hill. I have determined that these artificial persons can be controlled through their substantia nigra structure. The substantia nigra is located in the midbrain.

Similarly, encephalitis lethargica is a mysterious disease, almost forgotten by the medical community, literally and figuratively, and autopsies of people who have died of the high-mortality sleeping sickness have shown a deteriorated substantia nigra. So the nigra is used to control the movements of the androids on Capitol Hill and it is used to shut down (kill) people with a mysterious brain inflammation disease, a disease that essentially shuts down the operations of the body.

The substantia nigra connection is not a coincidence. It means that the nigra is some kind of 'control switch' for the android. Not just a control switch but also a 'shut down' switch, by controlling the dopamine mixture to the basal ganglia. The political puppets are being remote-controlled via their nigra switches. The encephalitic patients are being shut down, put to sleep and killed off by way of an *artificial program* (a symbiote for lack of a more scientific application) that is accessing their nigra structures (and likely other neurotransmitters and brain structures as necessary including the hypothalamus for sleep).

The kill off is a slow process, the body's movements are slowly turned off, the person is crippled physically, mentally, verbally, emotionally—all directly accessed via the nervous system. Often patients have psychological problems which could be reproduced by accessing the hippocampus (memory storage) and created new schisms and replaying traumas in the host. Once an intelligent organism has access to the command center, and learns how to operate the mental software, it can increasingly occupy the vehicle and use whatever means necessary to defeat or control the host, as it is programmed to operate.

While the strength and defence of the host plays a significant part in the ordeal (the disease) ultimately, since the host is biologically-minded and the disease is artificially-created, the host is overcome. Fifty percent of the cases during the epidemic were shut down, for example, they had respiratory failure. A large portion who survived were crippled or damaged and a significant portion remained asleep for decades, sleeping beauties.

People who were attacked by these shut down programs, usually on the first night of attack typically commented that they woke up "a different person." Of course it was different. It was different because they had been invaded by a symbiote, a created program. Who is creating these programs? Who understands androids? In short, the same people who created the androids. Why infect people with a disease? To cull the population is one reason. But more than that, it is a process to depopulate a certain class of people, an artificial class of people. The robot portion of a population. They need to cull the robots.

The question of culling robots, assuming we have come to some existential agreement that this is more or less accurate, is a more paradoxical question and not without answers. If there are robots living among us, if this is true enough, and if the world population functions on some spectrum of electromagnetic frequencies, then the frequency of robots should by all measure be distinct from the non-robots. Perhaps, the robot demographics functioned as a giant waveform of energy and this energy perhaps elevated the frequency of existence, or upheld it in some way.

In other words, life is better as a result of higher thinking. This waveform could have also acted like a firewall against technocratic pirates who had an agenda to bring war, turmoil, and suffering. In order to carry out those agendas they'd have to break the robot firewall. This could be one reason behind the encephalitis lethargica outbreak—the master controllers needed to take down the existential firewall; because what happened at the end of the twenties, the end of the outbreak? The stock market crashed and launched a global Depression. The Great Depression led to Hitler's rise and the devastation of WWII. Pretty good reasons to take out those damn robots. Of course, since we haven't been aware of the robots nor have we ever considered there were these advanced master controllers, we simply considered a mysterious illness that appeared and disappeared without rhyme or reason. I think mysteries have a reason for them occurring and I have briefly introduced my reason for the encephalitis lethargica outbreak.

If you are not a human-looking robot, or if you do not have a significant amount of synthetic DNA then you are more than likely immune (or are already turned off and not a threat). A class of person that is not artificial, or an artificial person with a more advanced operating system, will be immune to sleeping sickness, they might just get ADHD symptoms or they might become autistic.

As each person is equipped with an immune system, the advanced races need to compromise the immune system before the artificial programs can enter the host's system, latch onto the personal operating system and begin to attack their internal circuitry. In nearly every case of encephalitis lethargica, the patient at first had a case of influenza. In nearly every case this was true. The influenza weakened the immune system and allowed the viral symbiote to hack into the artificial DNA of the being. This is what the symbiote does, it overburdens the immune system with an artificial flu and when the immune system has been compromised enough then it can access the inner circuitry of the person. The result is quite disastrous—death, coma, cripple.

Encephalitis lethargica appeared mysteriously and disappeared mysteriously around the time of the Great Depression in America. Parkinson's disease today has no cause and no known cure. In each case, the substantia nigra has been damaged or isn't working properly. My own speculative additions to these scenarios included the following ideas: the people who got encephalitis lethargica were artificially created humans (androids) and the people with Parkinsonism have a significant amount of synthetic DNA (or mutated genes). In a synthetic person, access of the substantia nigra is access to the control center of the person. A symbiote accessing the nigra with its multidimensional tendrils, in the early days could shut down and kill the patient. A symbiote attacking a modern person, a person with a mixed genetic heritage or a more complex immune system, cannot easily put the target to sleep, but can damage the nigra structure enough so as to interrupt their signals to the basal ganglia, the switchboard to the physical network, and cause the symptoms of a post-encephalitic patient.

A person with Parkinson's had a defence mechanism to avoid death and to prevent coma but ultimately was invaded and then the midbrain was slowly damaged. All of this came about by a mysterious race of people who were involved in the creation of synthetic organisms. Who are still around. Who are still creating programs to incapacitate or kill people because diseases with no cures or causes are abundant (eg Alzheimer's, schizophrenia). But their evidence, like in the soft-titled "sleeping sickness," appears natural. It appears obvious. It appears biological. It isn't a sleeping sickness as much as it is a *biological hack.*

A heart attack appears biological. A flu pandemic appears biological. ADHD appears biological. Autism appears biological. Obsessive-Compulsive Disorder appears biological. What if all these diseases, and more, could be artificially created, given the right science? What if different programs were created by an advanced kind of programmer and those viral programs could hack into people (machines) and infect their hosts (through the nervous system) in any number of ways be it death, crippling or even identity control. Imagine that a symbiote could be created so as to attack the host in such a way that allowed the programmer to hijack the operations of the host, to even have the host perform actions that are against their nature. And then society would view those wild actions as insanity.

What if the whole idea of insanity is a deep misinterpretation of the artificial qualities of existence and the presence of malevolent programs and complex symbiotes? That would put psychiatry in a coma. I mean schizophrenia could be the result of a symbiote invasion on the host and the extremely unpredictable episodes were a result of the host (the person) trying to delete the symbiote program. It is a possibility. Of course, life isn't so simple and medical conditions are very delicate affairs because they also include psychological and spiritual parameters. The human persona is extremely complex and my observations, as wild as they are, they are still only scratching the surface of an entire being. What I believe we are seeing now is the multidimensional nature of life becoming much more prevalent and in need of deeper examination. It is a new kind of metaphysics, a science with a new potential for the human race. We haven't touched memories, programming, feelings, souls or other classes of beings. We haven't touched on the delicate sexual and traumatic factors. We certainly haven't answered all the questions and probably raised a myriad of new questions.

Did the Germans create a flu strain to hack into the artificial demographics and then accidentally killed all the other people as collateral because a virulent flu might affect different classes of beings differently? Perhaps they wanted to deactivate a certain class of robots (people with heavy doses of synthetic DNA). They release a dimensional weapon, a symbiote, the symbiote shuts down the host's operating system, puts them to *sleep* (much like a computer), a sleeping sickness. How do you deactivate a healthy robot (or computer)? You put them to sleep. How do you put them to sleep? You create a program that shuts down their operations.

I began this fringe medical investigation to explain the presence of advanced robots on Capitol Hill. I am satisfied that there is enough understanding to explain how these puppets are controlled—dopamine and melanin. I am confident of my observations of the three people I have seen others elsewhere as well. Most of all, the discussion on the mysterious encephalitis lethargica, its connection to the devastation of Parkinson's and that relationship to some kind of highly advanced artificial machine is extremely disturbing.

It appears urgent that we begin to shift our thinking from archaic biology to *artificiality*. That it now seems clear to me that the future of mankind, if it is to have a future, is to embrace its artificial roots and to admit the presence of an android class living among us.

ECONOMIC CRIMES

We are beginning to see the incredulity, the sheer technological magnitude of the situation and the kind of discussion that we have never been able to make. In my own view, it is unfathomable. Three androids were employed to hijack a nation. But the androids were programmed by their advanced makers, who we have yet to properly identify. And the bankers (eg Federal Reserve, US Treasury) were working in collusion with other government officials in the White House, the House and the Senate. And the military was involved with the production and release of a neurotoxin. And the pharmaceuticals were involved in the manufacture of a vaccine. And the intelligence agencies were involved probably at all levels. It's no longer just about the synthetic humans who commandeer America because, as we have seen, there are many elements to this dominion. It starts at the androids, at a place we have never considered before. We have lifted the veil of illusion of one of the most provocative situations ever to be imagined.

Once people make a commitment to the presence of androids in modern day they, at the same time, must recognize the mysterious android makers. This is a dimension of experience never before attempted. I think I have adequately explained, and documented by video, that there are at least three artificial persons (because there are many more) on Capitol Hill. Their builders can influence and coordinate all the parties mentioned into one of the most massive conspiracies ever imagined. And I think this will take quite some time to figure out completely, certainly I am not suggesting that I am doing that here, though I have brought to the light several key elements with a heavy-handed focus on the androids. I often struggled in the beginning with how much of the conspiratorial soliloquy I should expose, preferring myself to only focus on the androids. Each time I focused on the androids it was inevitable that a contextual situation came up, what they were speaking about, under what conditions they were deployed, and the more of those questions that I answered the larger the story became.

The economic crime itself—to hijack 311 million people—only justifies the employment of a level of technology to accomplish a very large task, a task that must leave no evidence. The fact that there was a financial crime is not as important as the fact that there are technologies 1,000 years ahead of humankind's best science. That is the real crime. It is an economic crime and a scientific crime. Imagine a robber goes to the bank using an advanced electromagnetic debit device where he walks into the bank and can transfer as much money as he wants into whatever bank account he wants and no one can trace him. He is not only stealing money, he is also using an advanced technology that the bank has no defence against. It is like raping a 1-year old child. The child has no chance. In this instance, you are the child.

The press conference took place in 2008. Congress wanted to stimulate the economy and bailout the failing banks. As you heard, this strategy was foolproof. It would create jobs and would prevent a recession or economic collapse. The general reaction was against a bank bailout and the $780 billion rescue. Conspiracy theorists said that the US had some outstanding loans with a secret society that if weren't paid, they'd be in trouble. So President Bush and the new President Obama became friends and agreed on the bailout, from the start, and then the android team was brought in to "seal the deal."

Could a person on drugs, a person suffering from drug side effects, a person with mental illness or neurological disorder manage a televised event? Could a heavily drugged or diseased person carry a press conference and, more importantly, would their colleagues entrust them to achieve a convincing job? Recall that these press conferences and speeches were very important, would you send a fool to handle an important press conference that would set in motion a series of extremely well planned events or would you send a human-looking political android?

The bailout was approved on October 1, 2008. A stimulus package of $700 billion (or so) was passed and the US President signed the bill just a few hours after it was passed on October 3rd. Now all the parties were happy. The public believed that the Wall Street Bailout would protect their jobs, save the economy and lower the deficit.

It happened at the auspicious time that one President (Republican) was transferring power to another President (Democrat). The audience felt relieved. The bankers were happy. The secret society would not have to do any dirty work. On the surface, all appeared quite normal. No one listened to the voices that opposed the financial architects.

We fast-forward two years later:

- Unemployment rate hit 10%
- U.S. Unemployment Figures: 2011 (8.8%); 2010 (10%); 2009 (9.3%); 2008 (5.8%); 2000 (4.0%)
- 2010 Budget deficit ($1.6 trillion)
- The economy was collapsing (fear of a second Depression was evident)

And we know that bankers control much of the world. The Federal Reserve Bank controls the flow of capital. So how is it possible that bankers saved America, prevented a Depression when the people's money saved the bankers?

More importantly, the very two items that the 2008 stimulus package was designed to protect—unemployment and economy—were at their worst. Is it possible that political policy proposed by Treasury Secretary Paulson, approved by President Bush, John McCain and President Obama and all their professional advisers on paid staff, could it be that they were all wrong? Were all these people incompetent and the people at the press conference mentally ill?

If I myself, having a poor understanding of finance and a weak mastery of economics, if I had given them economic advice, my poor advice would not have produced a worse outcome—a Second Depression. There was only one devastating Depression. Could their beautiful, perfect, ideal strategy to save America fail so horribly? The other question is—if the 2009 Depression was a result of *android influence* operating on an artificial network then were the same architects behind the 1929 Depression?

1929 depression

If the 2009 Depression was calculated and purposely created using the assistance of artificially created people, then it is likely that the 1929 Depression some 80 years earlier also involved artificially created people. Why is it likely?—because it has the same pattern. Same pattern, same people. Every professional, be it a thief, a military general and a filmmaker they have a basic pattern. A basic fingerprint. A *modus operandi*.

If the architects of the 2009 Depression used artificially created people, 'androids', then we can infer, and right now we don't know for sure, but we can infer that the same architects were present in 1929. The difference between 1929 and 2009, is about 80 years. So, the architects would have to be at minimum, 80 years old. That would make the architects 1-year old. Doesn't make sense, right?

Let's suggest that a gifted person could create a Depression grade economic shakedown, by age 30. If we choose an earlier age it might involve too much luck or too many advisers, let's say that a 30-year old could plausibly be the architect of a Great Depression. So 30 years + 80 years would make Depression architect 110 years old. He or she would be 110 years old planning this scenario for 2008, or alternatively teaching people to carry out their plan.

So we have a 110 year old architect working through artificially-created people who have control of the government, inserted into politics decades beforehand. These are very powerful people. But wait—we forgot about the androids in 1929. If the architect has a pattern today to do their dirty work then the architect must have also used androids in 1929. What kind of science did we have in 1929? First car radio, first phone booth, first color TV, first mass-produced sunglasses. We didn't have a computer until the 40s and the early computers used large tube transistors, filled entire office floors and didn't even have the power of a modern laptop. How is it possible that we had an android in the twenties? How is that possible? We could have made the argument that in 2008, having experimented in eugenics since the around the thirties, a subclass of humans, an elite culture that has found solace in secret underground installations, that they could have had sequestered certain scientific advancements, denying them to the public, and created these synthetic people. We could argue that they did this even in the late thirties to produce our three androids just be accident. But when you add the economic crime and the intent to hijack America you are no longer talking about a bunch of breakaway scientists. You are talking about a global agenda by people who have the means to carry this agenda out, in secret, through the activation of a human-looking class of androids. The encephalitis lethargica outbreak started in 1916, attacking another demographic of robot people who were at least 20 or 25 years old. That means that these people were engineered in the late 1800s and that means that eugenics was advanced enough at that time that they could create embryos in a test tube using synthetic DNA, even though DNA wasn't discovered until 1953. How could they manipulate synthetic DNA sequences in the 1800s, or any time before 1953, if scientists hadn't even discovered DNA?

Not only that, but, as I have said, and it is my estimate, the android technology of 2008 is about 1,000 years ahead of the best human science or robotics. Clearly it is very advanced, so advanced it has gone unnoticed by 7 billion people. It's ingenious actually. Quite impressive.

The three androids in my observation room were created in the early 40s. The woman is 71, the blinking man is 69 and the other man about the same. Their inception dates are smack in the middle of WWII. That also means that prior to 1940, there existed a technology to build an android and that science was at least 1,000 years ahead of humankind. Since no human had that technological capacity, we are dealing now with a mysterious race of people who can build androids and who can build human-looking androids and who have an interest in human leadership, and who are still at large. That is the other fundamental point; these technocratic masters are still very invested in global dominion.

This mysterious race of people is an advanced race. At least of 1,000 years ahead of humankind. Now they can be 10,000 years ahead of humankind, but they still have to build technology that fits into our level of civilization. They can be one million years more advanced, but still they have to build a fitting technology. If we were primates they would build primates. If they had four arms and we only have two arms, they would create people with two arms even though they could make someone with four arms. And if they are so advanced they probably do not exist in a temporal dimension so they can manipulate time to some extent, and I will bet that they can hide in the folds of space. They could be standing next to you but you cannot see them because they are vibrating at such a high rate that they are invisible.

Humankind is at Level X so they build Level X. Their capacity might be much longer. They might have the ability to build a metallic robot. They can build interdimensional ships. We can infer they are technological. But they have a deep interest in human progress. They are building politicians. They are programming and maintaining politicians so they are following or leading politics. They might have given us the idea of democracy so that they could insert their robots as our masters. They have an investment in human evolution. How long have they been investing (or interfering) in human progress?

About evolution: If these advanced scientists can create a human-looking android that can pass off as human, that can procreate, that can eat food, that can make friends and get drunk and they are able to do this in 1929, they might have been doing this for decades or centuries prior to this. It isn't possible that they landed in 1929 and created a Great Depression. The architect is 30 years old. The architect was born in 1899. He or she knows how to build and program an android. He or she might be physically 110 years old today. And he or she was the architect of the 2009 financial recession because the fingerprint is the same; therefore he or she is still alive.

Is the architect an android? To have the ability to artificially create a Great Depression, a strategic strike that would seriously harm the American nation, it would take an unusual set of knowledge, you would have to understand economics to a very high degree, you would need financial mastery, you would need political clout, you would need to coordinate with your androids, you would need to speak the language of your advanced scientific friends, people not of this dimension. You would need an agenda.

The architect is either an android or they are one of the advanced races. They are still alive. They are 110 years old, or more. And they can think on multiple chessboards at once. And they are mysterious. They are either invisible or they exist in their own dimensions. Or they may appear so normal that they are invisible. As I have suggested, the architects might be dressed up in an android vessel and living among us right now in some rural town in USA. They may take on several bodily forms. One of them might be a lovely 7-year-old girl that you would never think contains the consciousness of one of our slave masters. If we are to out imagine these masters, we at first, by default, have to catch up with the level of intelligence we are dealing with, and that level, as you have seen, is extremely high. But there is no escape from a prison that you do not understand.

Until I pointed out the androids we had no vocabulary for identifying them. So the architect is difficult to identify. The architect is 110 years old but may appear younger. Remember they are at least 1,000 years ahead of humankind. We can infer that if these advanced scientists can transfer human consciousness to an android body, an artificial body (or other body[7]) then it is highly likely that the architect could have transferred their consciousness to another body over the past 110 years. Perhaps more than once.

He or she could be 18 or 19 years old, or 25, and part of the British Royal Family. Remember, he or she is multidimensional. He can have a normal life but here and there can skip into other dimensions and put their plans into action. He or she, being infinitely smart and preferring wealth, would likely take the body of a royal class. You have instant wealth and privilege, privacy and political influence. You can hide in a royal family. You can also coordinate with other royal families and dynastic families and secret societies.

Also, if they can transfer human consciousness to an artificial body then they may have been doing this for centuries. They can be an immortal. Kind of like the immortals portrayed by Christopher Lambert in the 1986 fantasy film, *Highlander*. Lambert played Connor

---

[7] Assuming that they know how to transfer consciousness, they most likely understand how to incarnate, or reincarnate. They could, incarnate as a child. They could have chosen their parents, provided for them. Given them android acquaintances and then incarnated through one of them. Without a doubt, a race of advanced scientific people who can build robots that can procreate, that can download a consciousness into the robots; these scientists know how to transfer their own consciousness across a dimensional barrier and to be born as would a regular human. The difference is that they are infinitely more advanced, they are connected to their friends on the other side of the river (dimension) and they are perfectly disguised as a human being. No one would ever suspect it. And there is likely more than one. And they may have been doing this for quite a while. But if you stare at them as a human you will only see a human. This is not unlike the three androids I have identified. They have always been understood to be humans when they are not. So, there is quite likely a number of these advanced people in human form with a super advanced knowledge of reality, life and existential manipulation.

MacLeod, a modern day antiquities dealer, who had been alive since the 16th century. The opening narration sets the mood for the film: "From the Dawn of Time we came, moving silently down through the centuries. Living many secret lives. Struggling to reach the Time of the Gathering, when the few who remain will battle to the last. No one has ever known we were among you... until now." The film depicts an age-old battle between immortal warriors. Is that what it could mean to be an immortal—to be artificial? Are these immortals, androids posing as humans, traveling, thinking and influencing human history?

We have evidence that there is an android class of people. You have heard the basic medical evidence. There are these artificially created people and they were made by advanced races using technology at least 1,000 years ahead of human science. The architect of the 2008 orchestration is the same, or the same class, architect of 1929 because the pattern is the same—a flu, an economic plunge, a financial collapse. Why didn't a full Depression take effect by 2010? Was it because of the financial stimulus, as the architects suggest or is there some other factor? What if in 2010 there was a helping hand (same as in 1928 and 1945) and in 1929 there was no financial intervention. Perhaps there was another advanced race who prevented the worst case scenario.

Remember that there are androids on Capitol Hill that we have never seen before and all people involved in the 2008 Wall Street Bailout are part of a larger scheme perpetrated by an advanced invisible hand. The only kind of people who could counter that would have to be a people who were advanced enough to know about the androids. It seems logical. But humans didn't know. So who knew? And who took actions to reduce the damage? Somebody else is here now. Somebody who wasn't helping in 1929 or who failed in 1929; we don't know. Perhaps they weren't allowed to interfere before and now they have permission because otherwise humankind won't advance. Maybe it has been decided that humankind needs to progress this time around.

Another idea might be that the Architects of Doom achieved their intended result faster than in earlier times and eased back the destruction. Remember, we are technologically advanced and by clicking a few buttons on a software program, you can transfer trillions of dollars. In 1929, they didn't have computers, there was paperwork and that required time to process.

If there was a helping hand then that means there are other advanced races on the planet and they are helping humanity. For people who are familiar with my work, I can confirm that there are benevolent advanced people on the planet who are dedicated to human ascension and growth. But the android legacy is a new issue. It's never been discussed. No one explained this information to me.

Why in World War II?

There is something odd about creating android babies in 1940. That would mean conceiving the baby probably in 1939, at the start of the war. If the first android was accidental then why have two more in the next couple of years, all of it during a major world war. Our three android examples were born during World War II. Nancy Pelosi was born during the first year of the war. I have stated she is an android. Isn't there something strange about this?

In other words, at the worst time in human history they were manufacturing advanced beings. A major world war, military invasions and what was on their mind was creating robots. That raises a few flags. Because that doesn't make sense. Wouldn't these advanced people be focused on military strategies or training soldiers or developing military hardware? If they are at least 1,000 years ahead of human technology, why weren't they building laser tanks and teleportation pods? Why weren't they creating other advanced ideas like food pills or super soldiers?

They weren't focused on the war. Because they were building androids. They didn't want the war to end. If they wanted the war to end they'd have handed the Third Reich a squadron of scalar weapon tanks and antigrav jets. But they didn't. They built androids. Why? Because they didn't want the war to end. Because they created the war just like they created the Great Depression. Recall that there was an encephalitis lethargica outbreak near Vienna in 1916. That outbreak, a result of some weaponized toxin released during battle between Germans and French, didn't infect enough people fast enough. In 1918, war is quietening down and then there is another biological attack, H1N1 (same as in 2009). The flu is a virulent strain knocking people out, killing millions. Following that is the sleeping sickness which is a pandemic that lasts 15 years till about 1928. Then, not even a year later there is a stock market crash. Wall Street crashes (contrast this to the 2008 Bailout which still leads to a Depression) and there is a Depression. The Depression lasts 10 years until 1939 when Hitler and Mussolini are all lined up for a Second World War.

They took humanity from one World War straight into a pandemic then into a Depression and straight into a Second World War. If you cannot see the pattern of some very determined minds, you need to read this book again. You have the same virus, H1N1, present in 1918-19 and 2008-9. You have the same Wall Street in the United States. You have the same Federal Reserve Bank. You have the same substantia nigra. You have the same androids. You have the same architects.

Why else create a second world war? What better time to insert androids than a war. War time is full of confusion. No one is thinking straight. Orphaned kids are aplenty. They get new families. You don't know where they came from. The kids don't know who their parents are. The parents are too busy trying to survive to care. Perfect.

War, it's the perfect cover for an **android takeover**. You insert your androids, they disperse, they migrate, they become religious, they become educated, they grow up thinking they are human—But all the meanwhile, they are connected to our artificial android network that no one knows about because it's multidimensional and no one will believe in a multidimensional world until 2011; its science fiction. You have your actors in position for your next strike because that is all you do, you destroy, you interfere, and you kill. We are not dealing here with mass murderers. We are dealing here with extinction level exhibitionists. These performance artists will have you scratching your head as they wipe humans off the map; the entire map.

They have an agenda to control the world. They want people they can trust. They can program their androids to do whatever they say. They give these people a special card, special protection or inside knowledge. Enough of their androids survive amazing situations, for example, they survive the war by being in the right place at the right time, by moving to a quiet nation like Switzerland. They all get a special pass because they serve the puppet master and the puppet master is an advanced class of people who can predict the future and who are probably artificial themselves. Why are they artificial themselves? Because they are immortal. Because they have been entering and influencing human history for many centuries. Because they can inhabit other dimensions. Because they are technically advanced by at least 1,000 years.

What does it mean if the architects of the 2008 financial stimulus, which led to the 2009 economic catastrophe and the highest unemployment rate in the past many decades, are artificial themselves, that likelihood points to a new puzzle in the origins of the human species. It says to our most advanced geneticists and roboticists: "think synthetic."

THE ARTIFICIAL PEOPLE GRID

Outside of remote-controlled androids and historic crimes, the kind only imagined about in science fiction stories—and sci-fi filmmakers will have to up the ante of sci-fi films if they read this book—there is something else that I noticed that is happening. It relates to the outbreak starting in 1916 with encephalitis lethargica. It is by no exaggeration that the cause of the sleeping sickness was never discovered, that the strange attack of the autoimmune virus never understood and that there was no further explanation as to why it went away. A disease with no cause, no rational epidemiological pattern and no explanation as to why it stopped is not something that can be put aside without some kind of interpretation.

In reviewing the stats on encephalitis lethargica, we find the following: high deaths among infants, high deaths among 25-40 year olds, and high deaths among the 75-plus. What's the problem with this epidemiological pattern? Infants are born healthy, people in the peak of their lives are their healthiest, the only logical death occurrence here is old people who have weak immune systems and might die more easily. But that doesn't adequately cover the infants and the 25-40 year olds.

There is an additional hypothesis I'd like to add. I think it should be added because I think it might solve the mystery of encephalitis lethargica because I don't think encephalitis lethargica is encephalitis lethargica. Clearly, as you have read, it could have easily been created in a lab, it could have employed symbiotes, a large number of them, and the advanced scientists who created robots and inserted them into our world could indeed have taken the extra step to create a pandemic to kill off a bunch of people. But remember, these are advanced people, these are smart people, these people can foresee the future. They are not around to just kill people. They kill people for a reason. They can target people. They can design biological weapons to target specific groups (and likely individuals). Why target the healthy infants and the 25-40 year olds?

If the encephalitis lethargica pandemic was natural, in a natural viral infection environment, the weakest of the species get killed off. In this case, two demographic groups were the healthiest. From a medical perspective you can argue that the healthy immune systems overreacted to the flu strain, so much that it attacked itself and killed itself. That is a reasonable medical explanation as to why the 1918 flu killed so many healthy people. But that medical explanation does not include a) androids on Capitol Hill, and, b) advanced races of people. Neither do the typical medical experts include all the other out of the box stuff I've covered over the course of this book. So, the question returns to why kill off the healthiest people? If you wanted to kill people, you kill the weakest, you kill the poor, you kill off the malnourished, and you kill off certain races you don't like. But in this case, they targeted the healthy. This other factor, though covered briefly earlier in the book, needs a bit more exploration.

In any given civilization, it is probably logical to include a certain portion of artificial persons, whether they are androids or those made with synthetic cells, we don't know. We do know that people vibrate at different frequencies. We could infer that an artificial portion of a civilization would be required to be artificial in order to preserve a certain range of frequency in the global village, or to maintain a certain existential grid.

Recall, that we are now strongly agreeing with the fact that we are living in a technologically-created world because we have seen that androids are creatable beings. They are on TV. There are more of them elsewhere. So we also know that there are other people who have been artificially made. I think I can make the case that the infants and 25-40 year olds who died from the 1918 flu were artificially superior over the rest of the population. That means that an artificial "flu" was used to kill off android demographics. You kill off the infants, the new generation, you kill off the mid-generation, you kill off the old generation, and what have you accomplished? You have destroyed the essential artificial grid. A grid that could have been put in place to preserve peace, to preserve compassion and to protect against rampant corruption, destruction, war and disease. You have effectively reduced all generations of the android Diaspora.

Because that is what we have seen since WWI, haven't we? We now know possibly why the world is in constant chaos—they took out the artificial grid in 1918. They took out the primary defences between 1918-1919, during the Spanish Flu outbreak, and then finished the rest of the people off, those who survived the flu, with the symbiote adjustments. They shut people down (like computers), they shut down enough of the *artificial grid*, because we live in an artificial world. Madonna had it all wrong with she sang "we are living in a material world" because we are "living in a technological world." This also means that there is another race of scientists who are benign, who created the synthetic demographics with love only to watch them be killed off with some neurotoxin and symbiote devices.

At the end of 1929, perhaps the encephalitis lethargica-symbiote kill off program didn't run out, perhaps there was another advanced race that stepped in. And just as they stepped in, stopping the pandemic in around 1928, what happens? A Great Depression. One calamity after another. That is the signature of these advanced races: war, death, disease, suffering—they enjoy suffering. The Depression lasts 10 years and puts in motion the need for another Great War because the same people who ignited the First Great War are bent on igniting WWII, and they already have their actors on standby: Hitler, Hirohito, Mussolini, and Truman. And WWII stops after the atomic bomb. Why? Because another advanced race stepped in again. You can see the pattern here. People are so distanced from this kind of thinking, so afraid to step outside the box, to step into the science fiction world that they can't connect all the dots.

If anything from this fringe medical investigation, if anything, we have seen that there is an advanced hand trying to collapse the world and there is the saving hand (more advanced but not able to interfere) stepping in before it all caves in and rescuing the helpless people. And there are the people who are clueless.

What I have observed on TV, and tried to explain in this book is that there are advanced robots that look and act human and they are in leadership positions. There are others as well. They are not hiding in caves; these androids can be found as easily as watching television.

As long as we are sleepwalking through life, as long as we refuse to observe the way the world works, we will suffer endlessly. But if we gather our remaining drops of dopamine, if we turn on our magical powers with the melanin switches, we have a real good chance of preventing the next disaster and then preventing the disaster after that. Who knows, we could even break through the invisible barriers and catch these advanced scientists. We might finally stop acting like biological slaves and start acting like technological stewards.

Admittedly, it is a big thing to ask. It is a tall order. It means walking on the wild side. It means no longer living in the box. It means metaphysics. Some people like the box. Misery is a good excuse to get drunk, beat your wife and to take painkillers. I think we should change our focus and stand up to our technological masters who are playing us for fools. Because they have a list of crises and catastrophes lined up for us. Why they're creating new ways to make us suffer right now. I can hear their mental turbines grinding away.

INTERPRETATION OF OBSERVATIONS

We have stepped way outside the box of common knowledge and we did so in order to learn the secrets hidden in the universe. To some of you, I have made some very wild assertions and perhaps even wilder conclusions. To some others, many things I have said make sense even if not as clear as you would like. There are many things I understand but I am not a good translator. It is like I am Chinese with only a weak grasp of English. As an analogy, in Chinese, I can explain this very well, but not so well in English. Only in our case I am using a Reality Language and we only speak Orthodox Language. We speak Conventional Science, we speak Logic, we speak Math, I haven't met many who speak Reality. And no one in the history of this planet has ever spoken Android. Androids have never been so exposed as now. At least, this is what I think they are and the discourse remains solid even after now 5 years of android theory and research. In fact, the data I have been able to obtain in the last several months, some of it only suggested in this book, tells me that this work is far from over.

I think I have been able to crack open the android secret, a secret that was so secret only the most advanced (privileged) men and women know, and these men and women are nonbiological, they are artificial people who can manipulate time and space, they can program aspects of the world and they have been manipulating and influencing society with robots that look human. It is the kind of unimaginable scenario that has unfortunately been perfectly imagined. While we were being terrorized by invisible threats, rogue nuclear regimes, and evil alien invasions, during that entire chaotic disturbance, the real trick, the real magical mystery tour was happening right in front of us on live TV. The master illusionists had beaten us again.

A summarized list of my main arguments:

1) The United States of America has been hijacked by a very advanced kind of human-looking android under the direct control of their artificial makers;

2) There are three (3) artificially created (some mixture of synthetic DNA and genetically-engineered biological components) persons on Capitol Hill;

3) Their rapid eye movements and catatonic facial expressions are key medical clues pinpointing one of their methods of artificial control;

4) Their level of "life technology" is at least 1,000 years ahead of modern human science (technology);

5) They can be remote-controlled chiefly through their *substantia nigra* interface in their midbrain;

6) An advanced race of people, still unidentified and improperly understood, manufactured those androids and inserted them into society during the onset of World War II;

7) The 1929 Great Depression had the same architect(s) as the 2009 Depression;

8) A 15-year outbreak of Encephalitis Lethargica (and a deadly Spanish Flu) directly preceded the 1929 Depression and the Depression directly preceded WWII, and, a deadly outbreak of the H1N1 Flu directly preceded the 2009 Depression and the Depression preceded a new war in Libya, and pending wars (threats) in Syria and Iran;

9) The combination of the neurotransmitter dopamine with the dark pigment melanin forms a control switch to turn the android on and off;

10) A potent viral epidemic, an android leadership, a collapsed global economy, and warring nations are shared occurrences between 1929 and 2009;

11) There is a benign race of advanced scientists and technologists who might have stepped in to save humanity when at their lowest points, particularly in and around 1929, and they may have their own synthetic counterparts on the ground to prevent total collapse of the civilization;

12) America has been under the full administration of an advanced cosmic race, artificial machine-like intelligence, for (probably) twenty (20) years, perhaps more under lesser controls, and is not necessarily a free and democratic republic;

13) The android makers remain outside our perception, as did their androids until recently, and will not reveal themselves in fear of losing control of all their societal illusions (eg work, career, marriage, rights, government, theology, fame, wealth, etc.);

14) Many key politicians, business leaders, media personalities, celebrities, and influential thinkers have likely been duplicated, genetically-modified and engineered, or inserted as original androids, as a means to keep society sedate, distracted, and obedient, as well, so as to sideline any alternative ideas that might bring real change; and

15) The entire world population could be the result of a very old and ancient android lineage the likes of which are hard to imagine in a real world context.

If you want to know what is on their mind, if you want to know the thoughts of our unelected masters, identify their robots (and lesser minions), listen to what their robots are trying to convince your friends and families to do. If you do not agree, if you see the androids then support people who are against those deceptions; support the truth. In most cases, it isn't even necessary to explain who and what they are, rather it would make more sense to just empower society not to support those instruments of deceptions. The less support they have the harder they will try to convince society and the more they will need to expose themselves, until they are fully exposed. "You are not alone" is going to take on a whole new meaning. The android legacy on this planet is bigger than anything ever imagined. And that's what everyone says about their own great discovery... that is until the next great discovery.

My task—in the interpretation of the recorded observations—was to examine the real structure of the world, to overstep the usual selection of ideas, and to *bring forth* (by way of imagination) a new model of existence, one that has been properly and adequately updated, one that isn't perfect and complete and fluffy liken an Easter Bunny, which is what most curious minds would expect. The world has been made soft, our thinking is soft, if you think a little too left or a little too weird you're excluded from the normal crowd. The normal crowd, scientists included, is unflinchingly redundant and disassociated. It doesn't mind endless war, political corruption, childhood molestation, societal degradation, monetary consideration, newborn vaccinations; it puts aside progress (which requires imagination) for a careful sense of fluffy complacency. The normal crowd desires comfort and comforting words.

If you can take away anything of value from these series of intense observations, we must realize that we are blind to the presence of advanced technology because our very neurological processes are looped with fluffy expectation. We would rather stare at a large-breasted celebrity and vote on what she had for lunch last week than to notice the strangeness of a rapidly blinking high-level politician on prime time TV. People will criticize my discovery, almost certainly, but they will not criticize a celebrity who lost 7 pounds in 7 days. They will not criticize a new diet plan even though no diet plan has ever really guaranteed any results; the 30% obesity rates in North America should be a lot more convincing than the new diet salesman's pitch. On one hand, we have the capacity of observation and on the other hand we haven't the 'presence of will' to observe the truly important things in life. Humans seem to be afflicted with *aboulia*, a living paralysis. All I can hope is that people can see that this stuff is important. You need to think about this.

Perhaps humanity doesn't know what is important. Perhaps the decades of brain conditioning have essentially defeated the human will and the will no longer understands progress. A civilization, any civilization, progresses when it has within it the desire to progress. In having that will, the people are unafraid to imagine the future; the people are unafraid to imagine the future, to hallucinate a better tomorrow—isn't that what visionaries do? They hallucinate what tomorrow might look like. We call them visionaries. And yet, at the same time, the person who hallucinates something against the beliefs of a church or a national ideology, even if it is benign, is chastised, is hospitalized, is medicated, they are assassinated in any number of ways just because it is easier than saying *bon voyage.*

The three persons on Capitol Hill are actual living people. They have families. They have memories. They have beliefs and they carry ID. My observation of them has nothing to do with their lives or their identities, as I have stated. To me, who they are as individuals, as persons, is irrelevant. They are entitled to be a person, as each has chosen. But their artificial qualities, as I have observed and explored with my limited scientific abilities, pose a significant alternative life view. Their very presence, including their positions of power in the US government, indicates the fingerprint of a level of science 1,000 years ahead of a scientific genius on the planet. In statistical terms, it is an impossibility, isn't it? Androids simply cannot exist and yet I have explained that there are three androids on Capitol Hill.

If I am indeed correct, our entire world view is 1,000 years, or more, out of sync with the truth. The blinking woman is 72 years old. I don't care of age. Her technology is very advanced. That means, at the very least, in the best case scenario, we have been living in a biological lie for over seven decades. The last 72 years of human history were a waste of time, suffering, hope, and action. It was all one big gigantic lie.

It is a lie because there has been a technology 1,000 years (or more) ahead of human technology. It is a lie because artificially created humans have been directing human society. Not only that, but their programmers have used these mechanical dolls, these statuesque figurines to dictate what is important in life. And what appears to be important to these masters includes violence, war, tyranny, corruption, disease, hate, segregation, godlessness, negligence, neglect, disregard of the environment, dedication to wealth and achievement, vaccination of newborns, weather manipulation, cover-ups, defence budgets, secret societies, police states and the complete and utter repression of the human soul. And that might be it. The lack of humanity on the planet is because the planet is run by something that lacks humanity, mechanical puppets.

If the human race is to live as it should live, it should embrace an imaginative observation on TV and discuss it intelligently, not to vote, but to create a discourse on as many media as possible. And then people should take all manner of inconceivable topics—extraterrestrials, trees on Mars, secret space programs—and to replace the simple-minded fodder in media with things of substance. And that takes what, so far, society doesn't have—will power.

The argument isn't who is an android as much as who made the androids? Why did they make them? Where are they? Where are they from? What do they want? That's what interests me. It is like I just spotted a Boeing 747 in the year 1895. And I photographed the pilots and I showed the photos to a group of people and they began to believe me. And they wanted to capture the pilots and to capture the 747... But I would want to know who built the ship and how that 1969 aircraft made it into 1895, more than seven decades before it could ever be built because the technology wouldn't be available.

I think we should think about the presence of these beings and we should think about their creators. Their creators can answer all our questions. And their creators will not want to be found. Find them and we will find many answers about things we never even imagined.

COMMENTARY

THE ANDROID TAKEOVER

How many androids does it take to rule this planet? About one thousand. I could take over this planet with 1,000 human-looking synthetic people all of them programmed to serve me. That doesn't seem unreasonable. Is it achievable? Not with human science. Yet. At the conference in May 2010, pioneering microbiologist Craig J. Venter introduced us to the world's first synthetic cell and ultimately to synthetic DNA. Although many people may become overburdened with skepticism with the idea that key high-level politicians are programmable robots in the flesh, I certainly feel and think this is well within plausibility. The amendment I have made in order to derive this remarkable conclusion is the inclusion of advanced technological societies who are administering the human civilization.

Outside of human science are many other kinds of sciences including android science. A race of people so equipped with android science could create enough replicants to replace just half of the world leaders and to install an android-based administration system without any detection from the general populace. In fact, given the significance of my discovery on Capitol Hill, and especially given the fact that all three of my examples are high-level career politicians, I would have to infer that other nations have fallen under the same advanced protocols, and have been replaced accordingly.

There are 196 independent countries on this planet. From an external observation, this is a nice small number with which to execute my android takeover. I only have 196 targets. You see, the traditional wisdom, furnished through years and years of military propaganda, is that an advanced interstellar race of people would invade earth from the skies. That an armada of motherships and fighter ships would fill earth's skies in order to invade and takeover the planet. This is the conventional approach when it comes to evil alien invasion. The advanced intergalactic species wouldn't approach earth in any advanced way, rather they would jump in with both laser barrels firing. Of course, this is a childish notion that has been implanted into every mind that would buy it.

I like to think of advanced people as using their advanced thinking more efficiently. We have been led to believe that advanced threats would act in a humanistic fashion. We have been taught to fear a Martian invasion or an alien biological attack that would wipe out half the earth's population in seconds flat. But if we are able to extract ourselves from our own brainwashed minds, we would begin to realize, as I have, that advanced societies can take over a planet without ever a weapon being drawn, and I think that an android takeover is one of the best examples of an alien takeover if ever there was one. And all we would need is about 1,000 genetically-engineered humans.

Fifty-percent of world nations represent enough global influence to justify taking over their governing structures. If we round up the number of countries in the world to 200 then we'd only need to take over 100 of them in order to basically secure the earth. So, to start we would need to manufacture 100 robots, each of them tailored specifically to their nation. There are any number of ways to go about this. For example, we could outright replace the current leader with an exact clone of which is under our control. This wouldn't be too difficult to do as an advanced society. We might also create a brand new android and then mind control the current leader into obscurity. A simple political gaffe or inexcusable sexual innuendo should justify his or her removal from office. It would certainly prevent a re-election. Now that the current leader is on the way out that is when our political android just happens to rise to the occasion with a number of new inspirational policies and probably lower taxes for the Middle Class. Whatever the needs on the ground that is what we produce. We can program a person to fill in the shoes of a dictator and we can program someone to sing songs of democracy. The type of person is irrelevant. It's a matter of the programmers doing their job.

There is a better option. This one requires a bit more preplanning, but it works even better. What we do is we create several androids to fill in the key political parties, even competing parties. We don't care about competing parties. In fact, it is better if we have our ground troops playing all sides. In that way we can't lose. We create these androids early on, decades before, and then we keep them basically dormant for years and years. In the meantime we are playing with our more extravagant robots and our more robust thinkers, the kind of people who would lead a campaign to land a man on the moon, for example. Once we have flushed through these models, and once society has been disheartened and disappointed enough, that is when we take a look at our dormant stock and we think of activating one or two of them. I mean we might even have one of our robots assassinated in order to spearhead through a new kind of thinker or to further demoralize the human population. Humans are easily demoralized after their favorite leader is martyred. They remember and symbolize these people without knowing that they could have been androids, and they could have not been androids. Essentially, until human science is able to detect our synthetic DNA characteristics then we cannot be detected. The DNA of our synthetic models, for all intents and purposes, is the exact same as the regular human models. We would run into trouble in the future when the indigenous science is able to discern a synthetic cell from a non-synthetic cell. But by then the world is under our total dominion.

We start with 100 androids and take over 100 of the top nations. Once we have control of the government, and we have support of our beloved voters, it is then that we start to introduce our synthetic minions into the mix. We elect a synthetic defense secretary here and we position an android as the head of the intelligence agency. Again, these people can be cloned, they can be inserted or they can be activated from troops on the ground. Whatever works best. Ideally, we introduce several people at the same time who are competing for the same job to give the illusion that there is a choice. But there is no choice because they all work for us.

If we were to activate a total of 10 androids in each country, all of them in key positions of influence or power, we could very easily take the helm of humanity. We could influence the world in significant ways. We could set policies and laws that citizens would have to abide by. These policies and laws work in our favor and work against citizenry. We could establish scientific standards are that will never sufficiently allow humans to detect our presence. In other words, we could effectively cripple scientific progress. We could remove a few Teslas here and there, we could introduce fake scientific data to teach society what is acceptable and what isn't acceptable (adjusted to meet our needs for control, of course). We could say that certain herbs are hallucinogenic and therefore illegal. We could do the same for certain chemical compounds like LSD and make them highly restricted. We would want to curtail any avenue of human ascension and awareness progression.

We wouldn't want people taking LSD and seeing through the illusory structures that we put in place. We wouldn't want people to realize that the head of the government is a fully-functioning robot from the future. But we couldn't always do this in an authoritative way because people would naturally revolt and they would protest and that would hurt our ever-precious economy. That's why we'd have our robot lawyers sign new bills and laws that make certain things legal and certain things illegal. We would create disharmony in these substances by instigating and funding wars, for example, a Drug War. There is no need for a Drug War except to a) demoralize the people, and, b) to keep the rebels focused on the war and not what's really important. A war is a giant psychotic misdirection. A war is never a war, it is something else. It is a political manoeuvre, it is a distraction, it is a disguise for genocide—it is everything other than what it appears to be.

So now we are effectively able to cripple the nation with about ten robots. The more control we have the more time we have to activate and insert more of our robots, all of them looking human in every way imaginable. The female robots procreate. The male robots like to drink. Some are handsome and some are ugly. Some are smart and some are vulgar. Our created humans do not need to be perfect. Being perfect and beautiful might give their identities away. We later introduce some celebrity models and some art types and we are well on our way of a full takeover. Our androids then procreate and create a legacy of androids who all prefer our modality of thinking. We might lose a few along the way. We might have to sacrifice one or two here and there, but we can turn these tragedies into triumphs through our remote-controlled media outlets. Why the news anchors pretty soon belong to our invisible government.

As far-fetched as this science fiction scenario might appear to be on the surface, in my view, it is well within the realm of plausibility, especially given the androids I discovered on the Hill. There is no way that these three models are alone on this planet. No manufacturer would make only three models and then stop manufacturing. Plus, these three politicians are very old, hovering at age seventy. What does that mean? It means that the android manufacturers have had seven decades to manufacture and insert more androids. Not only that, but they have had the leisure with which to perfect their human replication technology to such an extent that the latest android models could be virtually undetectable.

The reason I was able to discern that Mr. Lieberman was acting strangely was chiefly because the master controllers were desperate to maintain their dominion over Americans. They needed to collapse the economy and therefore needed to convince the citizens that a financial rescue plan was an absolute necessity that would prevent the utter destruction of America as a nation. Because of their desperate attempts, they became sloppy. They overused the android features on the Lieberman model. That's why I was able to notice it. Sure, I had been paying particular attention to all kinds of anomalies since late 2005 and probably had a good eye for anomalous behavior, then again it could have been due that I didn't have much of a personal life and had nothing better to do than to study an old man blinking on prime time TV. In any case, my observations opened the door to the two other discoveries, and three anomalies spell the word "significant." One blinking robot can be dismissed as can two robots, but three robots—no way! Even I could not dismiss this as being significant. It turns out that there can be a benefit to having no personal life.

How many synthetic humans are in operation today around the world? Well, after more than 70 years, probably 100 years, and quite possibly for centuries, we can be certain that the key political, scientific, religious, educational, and medical posts are filled by synthetic puppets and are not necessarily in favor of the human race. We still don't know what these android makers want and that's because their androids were only really discovered in 2008 and not introduced publicly until August 2010. We won't be able to know what the android makers want until we can establish that my android theory is correct and that all these strange people I have seen since then are also synthetically derived. But the scientific plausibility, given Venter's amazing work in DNA science, is a non-discussion. And given my understanding of advanced cultures on or near the earth, speaking from direct experience, I can confirm that DNA science has been far more developed outside of human circles. In fact, many people who have had contact with interstellar races have often repeated the same kinds of stories involving genetic engineering and manipulation, even to read author Erich von Däniken writings that suggest the human race was genetically bred by an extraterrestrial race of beings.

There is no reason for any advanced race of people to invade our planet or to take it over through any form of military aggression. This is the kind of propaganda that has been effectively utilized since at least Halloween night in 1938 with the broadcast of the radio drama *The War of the Worlds* on CBS (Columbia Broadcasting System). Despite the flood of evil alien invasion scare tactics and brainwashing documentaries, a truly advanced and intelligent race would tend to act in an advanced fashion outside of any human capacity. It is very likely the case that many decades ago that enough of our people were technologically replaced with look-alikes and that we have already been invaded. Just no one told us about it yet.

MAKERS AND CONTROLLERS

The term "android" is misleading and does not always convey the exact nature of these synthetic humans. Similarly, the term "synthetic" does not properly register with human thinking. When we think of something synthetic we at the same time do not think of it as being alive, whereas when we think of an android we can also think of it as being alive. Synthetic humans are too new of a concept to make proper sense of and probably why my use of "android" was more appropriate. I think this is a fluid situation and over time people will have their own preferred words. Whether we refer to these creations as genetically-engineered people, synthetic humans, or human-looking robots is going to be a cultural decision as a reflection of scientific advancement. Right now I think the android descriptive highlights both the artificial nature of these people and, as well, it signifies that these people are alive; and these two things must co-exist in order for us to properly examine what has been carefully hidden from our sight for many, many centuries.

Oftentimes, a culture lacks the language with which to describe something and in doing so that thing sort of disappears from the radar. If you cannot adequately describe an event or experience, the human brain will store it away in a place where you cannot find it. We have seen this with young children who were sexually molested and how they have gaps in their youth, literal blank phases where they have no memory of what happened. It is because their experiences were so inexplicable and overwhelming, without diminishing the disturbing nature of these kinds of traumatic scenarios, that the brain naturally attempted to protect the individual from psychological harm. This is also the case with mind-bending observations that are too far from psychological understanding.

It wouldn't be unreasonable to think that because we lacked the language of a living android on prime time television that our brain decided that it was best to pretend that this scenario did not exist. It may have excluded this waveform of data when it processed the situation. When I first watched the white-haired man on the television, I saw him as anyone else saw him, but at the same time there was another layer of information. This sort of reminds me of looking at a painting, especially an artistic piece from a master. One person observes the painting and sees just the image. Another person, an *aficionado*, studies the painting and sees the beauty underneath the image. This was similar to the androids on screen.

After focusing my attention on the subject and putting aside the obvious I discovered another waveform of information. This waveform was heavily encoded and what seemed to be encrypted. The more I observed, and remained calm and open-minded, the more my brain began to decode the observation. I could understand why no one could see what I was seeing. I only began to see it through repeated observations. Each time I taught myself what any art expert learns to do, I taught myself to appreciate the visual beauty in front of my eyes. It truly had to come to this place of respect for the masterful work presented.

Truly, the geneticist who created these androids was a master creator. This geneticist was able to create life. These kinds of people aren't the random sort of people. These are masters who take pride in their work. If you study their creations close enough you would find their signature markings: certain skin tones, the height of the cheek bones, the disposition of the persona, the resonance of the voice, the vibration of the aura, the color of the pupils—all of these attributes were carefully manipulated in their genetic configuration. These examples were not paint-by-numbers androids, put them in a factory and pop them out by the hundreds. These were individual masterpieces that were carefully created with love, care, and a great attention to detail. In fact, I don't think any android life form can be created as coldly as is depicted in film. What the filmmakers truly lack is the possessive romance of a Dr. Victor Frankenstein, a

Swiss scientist who could not give life to his experiment
had he lacked his true love for what he did.

I often felt cheap or disrespectful when presenting this
information, especially when showing video clips of these
androids in action because it did not highlight the
mastery of these genetic artisans. By the same token, it
was like trying to explain Picasso to a bunch of
12-year-olds. There was no easy way about it and I chose
to highlight the android motif over the genetic artistry in
the hopes that people would one day recognize that these
synthetic humans were impressive creations that
demanded a certain amount of respect and admiration
(even though they were being presented as potential
threats to human sovereignty). I know what people must
think when I say things like this—we have to respect these
android puppets?—but yes, in a way, the geneticists are
not necessarily the master controllers. The geneticists in
question are those who have been 'hired' through some
means to create artificial puppets. They are not the
master planners of some world takeover. These are
artisans, like potters and carpenters. The carpenters just
build the house. The architects design the house. And the
billionaire who lives in the house, he might be the corrupt
individual unconnected to the builders. The carpenter has
no idea what the billionaire does in his house at night and
the carpenter doesn't necessarily care because he is busy
building yet another house. The billionaire might be
destroying the rainforest and ruthlessly vaccinating
African children and what he does and who this person is

may have nothing at all to do with the carpenters and architects of his house.

In this way we arrive at a general contradiction regarding this phenomenon found on video clips and television broadcasts. The maker of the androids does not have the same agenda as the owner of the androids, and the androids themselves, as individuals, are not individuals given free will and all the other stuff we normally associate with existence and divine rights. The androids belong to a new and distinct society that functions on multiple levels and dimensions, and these synthetic creations serve one or more nonhuman masters. These masters either exist outside of human perception, choosing to remain innocuous and therefore virtually invisible, or they physically exist in another dimension that does not require a bodily form, or they may exist inside of the android body itself. In any case, the master controllers have remained invisible to us, and so have their robot minions until now. If these androids are manufactured with a science that is roughly 1,000 years ahead of human science then it can be said that the makers and the controllers are probably several times that in advancement as cultures of people.

Why suggest that the makers and controllers are more advanced than their science? Because of the given situation. What is the given situation? There are human-looking androids that have been inserted into our modern society, at the highest levels of political power. But we forget one essential truth—the human civilization is still under development. In other words, the human life form, as an existential vessel, is not the most advanced life form necessarily. What that means is that the makers would not be interested in making a being that was truly advanced and fitted with all kinds of technological features. A car manufacturer, for example, hasn't the incentive or market demand to build an invisible car for the average suburban family. This does not mean that the technology and science is not available. It means that the market for cars requires them to remain visible at all times. This is the case for the human-looking androids.

It is likely the case that there are more advanced forms of body types, but they would not blend well in our society. In fact, say they were to insert an immortal being like Superman, well that wouldn't be as innocuous as expected. Human beings, leaning in favor of saviors, would likely come to worship this Superman android and that would destroy the intended societal manipulation. Even though they would have the technology to create more advanced forms of existence, it would be in their interest to create faithful simulacra since these would far better suit their needs. Their needs are to enslave and administer the human civilization.

They would not insert political robots for any other reason than to ensure that the course of America was in line with their intentions for all American citizens, and American citizens would blindly accept the course as dictated by their human-looking leadership, anything less starts to fall in the category of sedition. Then again, given the fact that the US Administration has been woefully hijacked by living androids who are under the control of nonhuman masters, it would seem that some kind of sedition is a requirement at this point. Already, at least two administrations, Bush and Obama, have been carefully manipulated by at least three robots, and likely more that we have yet to count. As explained earlier, there are quite likely many androids at the helm, even ones what we could not imagine were androids like say the US president.

What is the agenda of the controllers? Why have they inserted androids in a biological population? What do they ultimately want? Even though the controllers have not detailed their demands per se, we can infer what they want from us by examining their actions. As we discussed earlier, the fallout from the 1929 stock market crash led to World War II and within that chaos the controllers inserted more androids all without detection. We know this because we know the age of our three androids, they were born from 1939 to the early 1940s, smack dab in the middle of a global war.

Why would an advanced agency ignore the death of millions in favor of inserting more androids in the world? Because there is an agenda. The agenda says that these controllers want to remain in control and before they lose control they destabilize society with an economic crisis, they take away people's jobs and repossess their homes, and when people are demoralized and depressed enough, that is when they introduce a tyrant and a military trigger.

The trigger leads to a large-scale military battle. In the commotion of death, destruction, and demoralization, the advanced thinkers open the secret portals and release a new flock of androids. We, as distraught and suffering human beings at the mercy of disorder, haven't the resilience or foresight to look beyond our own need for food, shelter, and safety.

It is a rather impressive, if not sociopathic and evil, strategy. It is like going into the jungle, throwing in a few incendiary grenades, dropping a little napalm, and murdering any number of gorillas, all for the reason of destabilizing the collective and in order to clear a space so as to be able to introduce synthetic recreations of gorillas. Slowly, the gorillas recover from the shock and loss of their brethren and in the confusion they discover some new children. They unknowingly adopt these new children, as any parent or government agency would adopt an orphan of the war, without ever realizing their artificial nature. In fact, the trauma from the explosions and ensuing chaos have probably in some way altered the neurological wiring the gorilla brain, as a survival modality, and this too could have been a way to reset the brain's internal preferences.

It is true that sudden or extensive trauma, such as found in military engagements and invasions, do indeed reformat human thinking to some degree and prevent humans from ever accessing their latent potential (think of someone who has survived a mass shooting). And an advanced society that is capable of replicating a human, even making an improved genetic cousin, without question understands the human nervous system. They are playing us from Point A to Point B and back again, and preventing us from ever capitalizing on our divine gifts. Why is the android discovery important? Because it offers a chance to stop this enslavement pattern. It offers us a rare chance to identify our secret controllers and to expose their artificial minions, and in that sense it offers humanity its promise for salvation.

THE PELOSI PARADOX

Genetic artistry is a new kind of phenomenon that saturates the essence of the living androids, as it encompasses, without differentiation, every molecule of the living vessel. The appreciation for these existential instruments, as far as I'm concerned, is no different than the appreciation for a stringed instrument made by 17th century Italian craftsman Antonio Stradivari. A Stradivarius violin has fetched up to several million dollars at an auction despite the fact that their high valuations have never been justified, at least only subjectively, in any musical experiment. For many musical experts and players alike, the difference between a violin crafted by Stradivarius and one made by Greiner, a contemporary German violin-maker, is negligible. But for ambitious and competitive string players well-fitted on their career path, with or without the backing of a wealthy family, they strongly prefer the great old instruments.

Genetic crafts, as in art works, similarly rely on a subjective personification of market appreciation.

Currently, there is no auction for androids, mostly because there is no contemporary agreement that there are androids. The appreciation for any created instrument, whether it plays music at a concert or makes a speech on TV, falls upon the subjectivity of the observer. As much as my android discussion has tried to highlight methods to recognize that one type of human has been genetically engineered in a science lab, my aims ultimately cannot hold up against human subjectivity and the inability of any future observers to be able to discern the genetic genius in front of them.

In a matter of years, spending nearly each and every day on the subject, I have become a kind of amateur in android science, but an expert in android appraisals. It's an entirely new field and I haven't many, or any, competitors. Let's face it, I can make up whatever I want and it would be an expert observation. But this is not the method of any principled appraiser and certainly not in my best interest for a subject that is very close to my heart. Had someone told me when I was a teenager that I would become a global expert in android appraisals I would have laughed at the idea. Had that been a department of study at Simon Fraser University when I attended I might have even considered the idea, if nothing more than for the futuristic appeal. None of this of course took place. I chanced upon the living androids and they happened to capture my curiosity, and there went my time.

Lieberman, Baucus, and Pelosi, all born within 2 years of each other, represent three synthetic models; no different than talking about a cello, a bass, and a violin. They all have impressive qualities, most impressive of all is that these are living instruments that we call humans. Once we move beyond that weird idea, and it has taken me a number of years of full immersion to work that out so no one should expect to do it much faster than that, we get to appreciate these walking and talking instruments. All three of them, as a result of ageing, are now in their seventies and are considered old instruments. The most impressive android, of the three, is Nancy Pelosi. She represents the quality of a master craftsman. With no disrespect to the human modalities of Ms. Pelosi, and all of its accompaniments, her real beauty is found in her android features.

I cannot help but to isolate the Pelosi android as the Stradivarius of her generation. Aside from a striking woman, and putting aside her interests in cosmetic touch-ups, well within the American obsession for respect through beauty, Pelosi has truly been given a remarkable genetic heredity. The artisan must have been in a golden age of his or her craft when she was created in a lab. It may be difficult for an unscientific mind to think of lab creations as having any artistic value, but any endeavour can have artistic value, even a grocery clerk can take pride in their work. Genetics is just that much more removed from our daily struggle that we haven't the ability to appreciate all of its labors. But Pelosi is evidence that her craftsman indeed took pride in their work. She is a masterpiece. Of the three androids on Capitol Hill, and putting aside all others that have yet to be discussed, she is the premier edition. She is the great work. And I say it with all the respect for her as a human being, in fact, I don't discount the human being; but, the android—an entirely new phenomenon—is the real artwork.

I don't want the reader to think I am obsessed with Ms. Pelosi. I have seen some beautiful women, the Victoria Secret supermodels come to mind, there are many, and while I think that Ms. Pelosi is a beautiful older woman, having no access to images of her in her youth, what is most beautiful about her is her synthetic flourish. Pelosi the Android is a great old instrument that is far superior to all other instruments. Her music, no matter people like her or not, is enchanting from start to finish as a result of how she was made. Whoever made her, and from whichever lab she was made, was a genius. I can now appreciate that geneticist's genius. It may be that I am seeing these things because I am bored with staring at androids, or, it may be because I am learning to value these advanced creations living among us.

This is why I have also mentioned that, if anything, Ms. Pelosi is the likeliest candidate to embody one of the cosmic masters I have been searching for. Because of her refined molecular qualities, and unmatched DNA architecture, it was my guess that her creator, our master, is living through her body. He can do that because he understands that humans are basically experiential vessels, so he transfers his consciousness into her body in order to interact with the world. In his case, he embodies Pelosi in order to alter reality in his favor. With a world population untrained and uneducated in android strategies for global domination, Pelosi has come across as an arrogant and tough-spoken politician, and then this personality justifies her status as the most powerful woman in America. I don't dispute her political clout and I don't dispute the presence of her children, these have never been my interests. I have highlighted that she has another mode of operation and that happens to fall in the range of a robot. But now because she has been so perfectly engineered, and I don't mean that she is the most beautiful and intelligent, it would make sense that her engineer built her for a purpose—he wanted to embody her.

When you deal with living androids, you are dealing with two things, one, the body, and, two, the soul. They are two things that when put together make a person. In the advanced sciences these things can be pulled apart. They can remove a soul and they can replace a soul, and this is because they do not call it a soul, preferring to call it other things. They can transfer the ownership of a body just like a computer company can swap out the old operating system for a new version. If Pelosi's body is exceptional, it is because the occupant, the driver, is exceptional as well and obviously has chosen the Stradivarius over the Greiner. They know that genetics is everything and their first choice is a genetic masterpiece.

I think one of the things that has changed in all of this research is my appreciation of an advanced science. I started my quest with trying to find proof that these people were androids and then trying in vain to explain to the online public that my findings were correct. I have had only moderate success in this respect. The few people who agree with me have done so just because they can't raise a good enough argument that suggests I am flat wrong. Most people want to believe me but are afraid to go to that place so they await more corroborating evidence at some future date. A small number of people, even famous people, can see the androids on screen. They can see that I am serious and that my point is valid. They don't necessarily like what I've said, they certainly won't stick their necks out publicly, and they don't know what to do with this knowledge, but they are on my side (even if in secret). My own decision to accept my findings as they are and not to argue against the fundamentals, and only to reject some of the analysis, has taken me to a new level of android appreciation. Ms. Pelosi takes the first-prize in the android showdown. She has the kind of genomic vibration that beautifully renders the intention of her cosmic puppeteer. I think what impresses me the most is the fact that because of her brilliant shine I have, at least, found one of my cosmic culprits, hiding in plain sight at the highest levels of political power. There he is carefully tucked inside of the Pelosi droid. And I say this with the full understanding that most people do not believe in androids (yet) and, further, cannot understand how an outsider can embody another person.

LEGACY OF THE ANDROIDS

If it is relatively straightforward to create a human-looking android, one that can sleep, eat, and procreate, then in all likelihood human beings have just been given their overdue biological assessments. Given our analysis, and assumptions, it only would take a synthetic science advanced to roughly 1,000 years ahead of us today in order to perfectly create human replicants, even advanced replicants. And we can argue that there are competing existential platforms. But let's use our three political androids as examples of created humans. What does this do to the theory of human origins? If a synthetic science can create a living robot, do we still need an elusive divine maker? Are the android makers God? From a rational point of view, the people who made these three androids do indeed have the capacity to play God, and we have the recordings to prove that very capacity. But if this is the case, who created the human race? Did these same people create the human race? More importantly, if these three politicians are androids, artificial people that look perfectly human, then are biological humans not biological? In other words, are all human beings derived from a legacy of androids?

The android legacy argument was an intense argument I was forced to study in order to come to terms with all the things I had observed. Early on in my research, after an exhaustive approach to understanding the nature of the world, I determined, finally, that this world had been wholeheartedly manufactured. I concluded that we were living in a synthetic environment that only appeared real because we had forgotten how to appreciate its artificial qualities. As much success as I had with the external environment, I wasn't able to quickly come to terms with human beings. I could not explain how they fit into this nonbiological world if only to suggest that humans must be nonbiological. But I did not know how to express that in any meaningful way without just speaking of it as a belief. So I was stuck on that for a while.

The discovery of the androids happened in late 2008 with my understanding of the synthetic reality in fairly good form. I was convinced by then that the true nature of reality was synthetic. I suppose when I came upon Mr. Lieberman's unique disposition on TV, the blinking eyes and the strange patterns of movement, I was ready for the next stage of observation. I don't think I could've managed my observations had I not the synthetic foundation I had carefully developed. I was able to accept my intuitive response to these simulacra simply because I had already accepted an extended understanding of existence, and had theorized that humans were somehow nonbiological. As much as perhaps the external viewer might think that I saw Mr. Lieberman, having done nothing else with my life, and then jumped up screaming "Android, android!" Clearly, this is not the case and I have never implied it is the case, but perhaps due to a limitation of space or circumstance I have not always explained the full story and not everyone has read all of my printed works which tend to favor back stories.

Oftentimes, even documentaries cannot fully convey the historical approach, and oftentimes the reason for this is that the material is quite boring and not as relevant to the medium as the data and discussion. So my discovery happened after at least three years of intense research and then it still took me another 18 months or so to process and decode at least some of that data, so we are talking now about 5 years of work and preparation *before* any public announcement on artificial people. And even then the discussion didn't gain clarity until I further educated myself and improved on my theories over the course of another year. At the time of this writing, it is now the fifth year after my discovery in 2008. As you can see there is much to discuss and there is a need to continually update my information. In fact, there are so many androids I have seen now that new books are required and that this investigation must stand on its own. I foresee that in the years and decades to come we will be able to build a library of knowledge regarding android science. It is obvious to me that as impressive and difficult this work has been, and the toll it has taken on my personal life, that it will ultimately reveal itself to be an introduction to a new branch of thinking.

This branch of thinking will invariably cross a number of other disciplines, specifically human origins. My approach to material is to keep things simple. If these three politicians are synthetic humans then the science and know-how is available to make "people." We can assume that these "people-makers" have been around longer than say a few thousand years and therefore could have very well created some of our ancestors. How far back did they start creating? We don't know at this point. We do know that humans can be synthetically created, programmed, and controlled. This is certain. We can infer that all humans have been created synthetically to some degree or other, although it must be stipulated and often not enough, that the technology to create humans is extremely advanced, so advanced that it appears natural and biological. We could call it divine technology, and it is something still far beyond our comprehension.

Even though I feel it is unfair to write about it in such a pedestrian fashion, I cannot help but attempt to explain that there are things that are far beyond our comprehension in use today. To properly explain these cosmic instruments would require a different kind of technical language, which we have not acquired thus far, or it could be appropriately explained using the kinds of terminology commonly found in biblical scripture, which I think would only confound the issues at hand; therefore, it is best to keep things simple until such time that people are ready for complexity.

More than that, as my recent work in DNA has revealed, complexity is found in simplicity, which tells us that the more we learn to observe the obvious the more we understand complex things, rather than the other way around. We have been educated to hear a complex explanation and to determine that the idea is complex when in fact this, if my research is so far correct, isn't exactly true. The more simple the explanation the more complexity it contains. I think that this fits well with my observations of three regular people who all turned out to be people of an extremely advanced nature, and didn't need to turn into titanium alloy mechanical parts, as we would normally expect in cases of robots and computers. That the computer is made of molecules is a wonderful alternative to a computer made of transistors. We just need to realize this.

The creation of humans and synthetic environments ultimately falls into my pioneering work on reality science whereby I have been able, with some luck, to detail a new story about life on earth. My technologically-based story does not require religious belief and it does not require extraterrestrials, it only requires the availability of a cosmic science. I am certain that this cosmic science is available. I have seen the results of this science on screen, as have well over 40,000 viewers online. The secret is out.

TECHNOLOGICAL TYRANTS AT LARGE

At some point we have to face the situation regarding the presence of androids... at some point... and to make some hard decisions regarding their influence. Nancy Pelosi served under Presidents George W. Bush and Barack Obama. Pelosi is one of my three androids that I discovered. She is clearly out of sync with reality during the State of the Union speeches, not just under Bush but also under Obama. On camera, I have recorded her in android format—whether blinking wildly or staring eternally and it is very clear even to an outside observer that Mrs. Pelosi, with all due respect, is simply not functioning within the human life spectrum. So what do we do? Do we choose to ignore the fact that the highest level US female politician is a robot or do we begin to think of this as a major threat to American sovereignty? I'm not even American, but I am deeply concerned, so concerned in fact that I have produced several video documentaries and published books on the subject.

After two years of speaking up and making presentations, I can say that this topic is trapped in people's minds. The few who sort of believe me are struggling with internal conflict and the many who'd like to believe me haven't a clue what to do with this revelation. Here we are in a society that demands the truth, with people who want to ascend, and here is a well-documented discovery including extensive video evidence and even improvised pseudo-scientific analysis, and besides me, where are the champions? Nowhere to be found. Again, the most powerful woman in America, third in line to run the country if the President and Vice-President were indisposed, who isn't a natural born human. She was configured in a secret lab and by all guesses was conceived in a test tube. More so, Mrs. Pelosi, by my professional analysis, is a living android. She's not only a synthetic robot, but she serves another master. She serves some entity that humanity has yet to face. She doesn't serve America because she is not American. She's a robot disguised as an American. We don't know the full agenda of her masters but it is safe to say that the dark agenda is not in humanity's best interest. There's no way around the inevitable—we need a larger investigation and greater exposure into these matters.

We have learned of a method of societal administration that is invisible and undetectable. And even the discovery of this method has not ensured the return to the sovereignty of man, we are thusly not much better off than we were when we started. A truly free people, in which people had regular access to various forms of classified and alternative information, would naturally have defense mechanisms that would prevent a societal hijacking. If the history of coup d'états reveal anything, in the traditional sense, the administrative procedures of a nation, and the collective will of the people, decide the exact nature and tradition of a nation. By definition, a leadership, whether naturally voted into office, militarily rebooted, or, now, invisibly inserted without detection would still be counted as a leadership and its citizens would adhere to the demands of the constitution and laws. In other words, even a hijacked government can hide behind constitutional legalese and once they are firmly rooted in place the people would have to apply the laws if they wanted to remove them. This is especially true for developed nations such as America.

Furthermore, proving the synthetic nature of any of these highlighted individuals, given our current access to technological instruments, is probably unrealistic, though observation clearly communicates the necessity for investigation and concern. Removing high-level politicians from office, if the attempted impeachment of Bill Clinton is any example, is extremely difficult. And given the pervasive nature of this android hierarchy, not withholding the systemic corruption, and the advanced characteristics of the controllers, any attempt to remove the androids and their minions would require an equally advanced strategic approach. This line of thinking is essential to anyone who wishes to utilize this speculative analysis to the betterment of humankind.

There is no need for us to magnify the situation we are presented with and to do something that is in itself irresponsible and simple-minded. We need only to start with the very basics of interpretation and discovery, that America, as the most powerful nation in the world, is under the dominion, and has been under the dominion, of a very advanced scientific culture. This advanced culture is also behind the mass vaccination programs, the mass distribution of global aerosols (chemtrails) and ecological transmutation, the cover-up in space exploration and moon habitation, and the genetic modification of human food (GMO food), just to name a few examples of global interference. The pervasive intrusion into the natural and harmonious way of human existence has led to the destabilization of the planet and has continued the millennia of blood offerings (now disguised as war for freedom) and rituals from long ago.

But there is a problem. The exceptional presence of android cultures in America, which were only presented publicly for the first time in 2010, and which have been only considered a possibility in science fiction works, does not easily lend itself to an expedited solution. The conspiratorial features indicated by the inclusion of political, financial, military, and medical perspectives, provide fierce barriers against their eventual removal. Worse still, with the current level of human knowledge completely devoid of such eccentric observations, the period of time required for a full and proper education on the subject would only allow the advanced agents more time to implement even more extreme enslavement protocols. By the time that people started to accept my hypothesis, that we are under an android administration, we would have all been vaccinated with gene-stopping vaccines and microchipped and hooked into some quasi-mind altering network interface. In one or two years, there could be immense societal shifts as a result of more technological activation, including the implementation of round-the-clock drone observation units across all parts of the nation. On the one hand, we cannot afford the time it takes to inform the public that there are artificial people in control, but on the other hand, and if history teaches us anything, people need time to process and absorb this kind of ultra-fantastical thinking, even if it is true.

In any case, the solution to this problem will have to be improvised since there never has been this kind of scenario on earth ever before. That should not sink the aspirations of the freedom fighters, although it will certainly deter anyone with a good reputation, and neither should it exclude those who are less than proficient with android identification. Similarly, attacking the androids serves almost no purpose because these are computer terminals, the controllers and the programmers are elsewhere, making more androids, and as long as they are elsewhere then humanity is going nowhere, and that is because the controllers do not want humanity to wake up. If this discovery tells us anything then it tells us that there is an invisible government, still not delineated, that is in control of most of this planet, not all of this planet, but certainly the key functional areas, the major cities, the major nations, and their chief institutions. If there are 196 countries on the earth and 193 of them are members of the United Nations, we can assume that controlling the UN would enable an agency to virtually control the world, and therefore if you want to find more androids, go to the UN. In fact, there may even be secret UN departments that are designing, programming, and manufacturing robots. It's just a matter of exploring all of your imaginative faculties, as I have demonstrated here in my own imaginative investigation.

There is no easy way forward, and we have to avoid any form of regression. This is what the controllers want. They want human regression. Humans naturally desire progression. And so this is the battlefield. If we want progress then we will without question, and with a few scars, have to rise to the challenge of our wardens. They have every advantage that an incumbent president has and we have all the disadvantages of low self-esteem and the memory of a person with Alzheimer's, so those people who are not up to this kind of thinking should be asked to step aside and in their place we allow the real heroes to step forward.

It remains to be seen what will transpire, but what will transpire is what has probably always transpired when one culture had been imprisoned by a more advanced culture. The emancipation from oppression, in all its forms and formats, leads to the establishment of the sovereign individual, and the collective of sovereign individuals is the basis for a sovereign culture, and a sovereign reality. Any agencies and persons not welcome in our reality system needs to be appropriately deported with a sense of haste once they have been properly identified and examined. Round them up and ship them out, once and for all. Earth deserves its sovereignty back. Take back your planet from your android oppressors before it is too late.

APPENDIX

AN UNSCIENTIFIC BLINKING TEST

The initial observation noted the unusual case of rapid eye blinking. While a saccadic eye movement and the substantia nigra activation are good medical explanations, from such a superficial observation, I think that this has caused quite a bit of confusion among many viewers. It seems that a lot of people are fixated on the eye blinking when in fact the eye blinking is but one of the simpler observational phenomena.

As a kind of late practice for the game, I decided to do my own blinking experiment—to see for myself if there was any unscientific evidence to be gleaned from the blinking. Till now, I had relied on my own observations. Joe Lieberman was blinking unusually fast. His blinks were rhythmic and mechanical. Nancy Pelosi as well. Max Baucus was blinking much slower than normal, unusually so. Those were my basic observations and statements, but I did not provide any data. So here I am going to do a bit of (casual) data analysis on the blinking phenomenon. Again, I need to reiterate that the rapid blinking is just one of the aspects of my immediate medical investigation (and represents a key trait only found on certain models). To me, it is like noticing an odd curvature, or a strange symbol, on the wall and then with further inspection you discover a secret door and that secret door leads you to a lost tomb. The discovery is not the odd curvature nor is it the strange symbol. Sadly, this is what many people are focused on. The discovery, in my opinion, is the lost tomb. In this case, the discovery is the three androids on Capitol Hill.

Perception is everything in the alternative business. If you focus on the wrong aspects you end in the wrong conclusions. I have seen this before in regards to prophetic statements—that the re-reading of biblical prophecy coupled with some astrological phenomenon can lead to some new end of the world scenario. But the combination of some biblical statement translated, written and edited 2,000 years ago with some modern day discovery of a comet flying through space at a certain trajectory need not be related. Oftentimes these kinds of unrelated items and even individuals (eg Presidents) get combined together into some new apocalyptic event.

To focus on the blinking as a determinate for the presence of an *android* is misguided and simply wrong. It is not the way I have done my work, but I have used the blinking as an easy-to-identify indicator so that my audience would not fall off had I presented even more advanced observations (eg tactical positions, blinking during certain crises or messages). Nevertheless, I think I should finish up the blinking observation with some data. The data presents additional support to my initial observations which were fairly accurate. The blinking has even other characteristics which I noted but I will not use since that would cause more confusion and add to the confusion that already exists. I think this could be discussed in later years.

I decided to use eleven subjects as my control sample. I handpicked people who were otherwise stable and had a consistent reputation, no known drug addictions and no outright strange behaviour. My control group had to be TV personalities or professionals. I needed to match this with the politicians, all public speakers, experts on camera and in speeches.

Prior to this data, I had never counted the blinking rate. I chose to start with my control group and later count the blinks on my three subjects. I therefore didn't know the benchmark. I had no bias on the blinking speed. As you have probably read, the blinking rate I attributed to some kind of "data transmission." A faster blinking rate was equivalent to a faster rate of transmission so that when an android blinks rapidly there is some engagement with another person or an audience. A slower rate could be attributed to a different kind of transmission or even perhaps a kind of download.

Of my eleven subjects, I had selected 6 men and 5 women. At first I would focus on the men but then decided to have some women since I had a blinking woman in my discovery. I probably should have used three or four women since that would be in direct proportion to my sample, two men and one woman (33%). In any case, I found people relatively similar with only a few descriptive differences as you will see.

The lowest rate of blinking was found in two men. They were blinking at a rate of 8 blinks per minute. I tried measuring blinks both using a pen and paper and, later, just by observation and count. I used a timer to provide a reference and then compared the lapse of time and blinks to determine a rate. So this isn't perfectly accurate because sometimes the speaker wasn't on-air 100% of the time and other times they were too far from the camera to see their eyes clearly. I had to improvise and guesstimate. Luckily, I could view the same people in different shows or episodes and then could average out the differences. This removed many of the errors.

The women tended to blink faster and I attributed this to their emotional state. Women appeared to wear their hearts on their sleeves, so to speak, so that their emotional state improved the rate of blinking. Being on-air also stimulated those brain chemicals like dopamine and that too likely increased the rate of blinking. It wasn't strange then to notice that the women tended to have a higher blink rate as compared to the men. The highest rate was 43 blinks per minute.

I average these out and matched them against some of the more common figures between 26-35 blinks per minute to get a rate of about 30 blinks per minute. I think this was a fair rate. Now I put this 30 blinks per minute and put it to the side.

I started to view the three alleged androids, starting with Max Baucus. His lowest rate of blinking was 4 blinks per minute. On average he was blinking 8-10 blinks per minute. I used the average of nine. Also I noticed something else—a person doesn't blink in a rhythmic fashion. In other words, they don't blink once every two seconds (30 blinks per minute). Rather they might not blink for 10 seconds and then blink four times. Over the course of a minute we can get an average rate. Baucus, at times, did not blink for 55 seconds and then for those last 5 seconds blinked 5 times. Although I used the 5 blinks per minute rate, it wasn't a true statement. All of this I figured into the data and my extrapolation of it. I couldn't simply rely on one observational method.

I moved on to Joe Lieberman, the initial observation, clearly the fastest blinker of them all. Now I was to find out how truly fast he was. His rate was very high. He averaged 119 blinks per minute. Let's round it to 120. He was 400% faster than the average of the control group. He was blinking twice a second! And he did it for nearly all the 11 minutes. That is the more incredible aspect. He was not only blinking rapidly, but he held it for a prolonged period. In contrast, my control subjects might blink at a higher rate for several minutes. I did not see them sustain it for any particular length.

Pelosi then presented another important observation: the *State of the Union* address was 50-minutes to 70-minutes in length, and Nancy Pelosi maintained her blinking throughout. Her rate was between 50-70 blinks per

minute, or 60 blinks per minute average. That is still
200% faster than the control group (30 blinks/min). The
control group, mind you, consisted of normal people so
that to me meant that my androids, at least two of them,
were beyond normal (or abnormal).

So, at his slowest point, Baucus was 50% slower (4
blinks/min) than the slowest of the control group (8
blinks/min), and about 70% slower than the average (9
blinks/min vs 30 blinks/min). Indeed, he is significantly
slower.

The rapid dopamine blinkers were significantly faster
than the control group. If we took the fastest *normal*
blinker (43 blinks/min) and compared her to Lieberman
(120 blinks/min) then he is nearly 300% faster. And the
blinking female Pelosi, at 60 blinks/min, is still 25%
faster and, more importantly, was sustained for an hour
or more.

I randomly selected my control group using some basic
criteria. I established their blinking average beforehand
and then compared that to data I obtained from my three
synthetic individuals. Overall, all three of them were
exceptional and I think further supports that the blinking
indeed is related to their synthetic characteristics, for one,
and two, that this blinking as a rate of transmission is
likely the case. But the rate and consistency of blinking
does not alone indicate they are androids. In other words,
a slow blinker or a fast blinker are not necessarily

androids. This would have to be weighed in with all the other characteristics and observations.

As well, blinking was influenced by emotional state, level of anxiety and even dryness of the eyes and amount of wind. Some of the control group blinked faster if they were recalling information so we could say that memory recall and blinking were related.

20399511R00134

Made in the USA
Charleston, SC
09 July 2013